1695 EM

Phase Equilibrium in Process Design

PHASE EQUILIBRIUM IN PROCESS DESIGN

HAROLD R. NULL

Engineering Fellow
Central Engineering Department
Monsanto Company

WILEY-INTERSCIENCE

a Division of John Wiley & Sons, Inc.
New York · London · Sydney · Toronto

CHEMISTRY

Copyright © 1970, by John Wiley & Sons, Inc.

All rights reserved. No part of this book may be reproduced by any means, nor transmitted, nor translated into a machine language without the written permission of the publisher.

Library of Congress Catalog Card Number: 70-130432

ISBN 0 471 65185-0

Printed in the United States of America

10 9 8 7 6 5 4 3 2 1

To My Wife June
and children Jody, David and Debra

PREFACE

The literature on phase equilibrium is voluminous, and several excellent texts have been written on the subject. As a result, the engineer or scientist intending to specialize in the subject has at his disposal a wealth of reference material. I have found that, in spite of the vast literature, the subject is poorly understood by the engineers having responsibility for the design and development of chemical processes. This situation is especially disheartening because the success or failure of most separation processes depends on a favorable equilibrium relationship between the phases, regardless of the high efficiency of the equipment.

The lack of understanding of phase equilibrium among process engineers is due, I believe, to the tendency of authors to write primarily for the specialist, who wishes to explore the subject in depth and with mathematical rigor. The design or development engineer cannot become a specialist in phase equilibrium; too many other subjects demand his time for a disproportionate share to be allotted to this one discipline, however important it may be.

In this book I hope to provide a text and reference that meet the needs of design and development engineers. As such this volume should serve as a text on the senior or first-year-graduate level, or as a text for continuing education of practicing engineers. Although a certain amount of mathematical derivation is necessary to the understanding of the principles, I have attempted to use derivations only where they enhance the understanding of a principle. Where I have felt that a sacrifice of mathematical rigor contributes to the clarity of the presentation, I have taken that liberty. I make no apology for this, since the understanding and use of the material presented are more important than a demonstration of mathematical manipulative prowess.

It is inevitable that a book such as this should emphasize the items that are most familiar in the experience of the author. In my work in engineering research and process development, and as advisor to a number of design engineers, I have found that certain techniques are particularly valuable in expediting the quantitative representation of the phase equilibria involved. Some of these techniques are theoretically derivable, whereas others are largely empirical. I have attempted to present these as clearly as possible in the hope that they will prove as useful to the reader as they have to me.

I am indebted to many persons in the preparation of this work. The many research scientists and engineers who have published in this field are obviously contributors; I wish to state in advance that I have not attempted the impossible task of referring to them all in the text. More specifically, I am indebted to the management of the Monsanto Company—particularly R. W. Schuler, J. B. Duncan, J. R. Fair, and H. W. Martin—for encouraging this work and for releasing some of the material that was developed in the course of my work in the Monsanto laboratories. I am also indebted to several of my colleagues who have reviewed the manuscript critically. Lynn Bellamy and David Palmer have been especially helpful in this regard.

I hope the material contained herein will help to fill the gap that I feel exists between the phase-equilibrium specialist and the practicing engineer.

HAROLD R. NULL

St. Louis, Missouri
April 1970

CONTENTS

1 **Introduction**   1

2 **Basic Thermodynamics of Phase Equilibrium**   3

    2.1 Introduction, 3
    2.2 Important Thermodynamic Functions, 4
    2.3 Thermodynamic Conditions for Phase Equilibrium, 5
    2.4 The Gibbs–Duhem Equations, 7
    2.5 Mathematical Relationships between Phases in Equilibrium, 9
    2.6 Ideal and Regular Solutions, 11
    2.7 Fugacity, 12
    2.8 Summary, 15

3 **Evaluation of Standard-State Free-Energy Difference**   17

    3.1 Introduction, 17
    3.2 Vapor–Liquid Equilibrium, 18
    3.3 Vapor–Solid Equilibrium, 21
    3.4 Solid–Liquid Equilibrium, 21
    3.5 Liquid–Liquid Equilibrium, 23
    3.6 Solid–Solid Equilibrium, 24
    3.7 Gas–Liquid Equilibrium, 24
    3.8 Summary, 26

4 **Equations for Activity Coefficients**   28

    4.1 Introduction, 28
    4.2 Gas-Phase Solutions, 28
    4.3 Liquid-Phase Activity Coefficients, 37
        4.3.1 Two-Constant Binary Equations, 40

        4.3.2 Three-Constant Binary Equations, 44
        4.3.3 Multicomponent Equations, 48
        4.3.4 Correlations for Infinite-Dilution Activity Coefficients, 55
  4.4 Solid-Phase Activity Coefficients, 61
  4.5 Summary, 65

5 Vapor–Liquid Equilibrium   68

  5.1 Introduction, 68
  5.2 Types of Phase Diagram in Vapor–Liquid Systems, 68
        5.2.1 Ideal Solutions, 69
        5.2.2 Nonideal Systems, Nonazeotroping, 71
        5.2.3 Low-Boiling Azeotropic Systems, 71
        5.2.4 High-Boiling Azeotropic Systems, 74
        5.2.5 Heterogeneous Azeotropic Systems, 74
        5.2.6 Heterogeneous, Nonazeotroping Systems, 77
        5.2.7 Thermodynamic Interpretation, 77
        5.2.8 High Pressures, 82
        5.2.9 Multicomponent Systems, 86
  5.3 Experimental Vapor–Liquid Equilibrium Data, 90
        5.3.1 Single-Stage Vapor–Liquid Equilibrium Apparatus, 90
        5.3.2 Multistage Apparatus, 93
  5.4 Processing Vapor–Liquid Equilibrium Data, 97
        5.4.1 Data on Pressure, Temperature, Vapor Composition and Liquid Composition, 99
  5.5 Process Calculations, 145
        5.5.1 Bubble-Point Calculations, 146
        5.5.2 Dew-Point Calculations, 164
        5.5.3 Flash Calculations, 171
        5.5.4 Complete Vapor–Liquid Equilibrium Phase Diagram, 174
        5.5.5 Process Applications, 174

6 Liquid–Liquid Equilibrium   183

  6.1 Introduction, 183
  6.2 Types of Phase Diagram, 184
        6.2.1 Binary Systems, 184
        6.2.2 Ternary Systems, 186
        6.2.3 Multicomponent Systems, 189
        6.2.4 Effect of Pressure, 190
        6.2.5 Thermodynamic Interpretation, 190
  6.3 Experimental Liquid–Liquid Equilibrium Data, 192
  6.4 Processing Experimental Data for Liquid–Liquid Equilibria, 193

Contents

      6.4.1   Binary Systems, 193
      6.4.2   Multicomponent Systems, 204
  6.5 Process Calculations, 205
      6.5.1   Binary Systems, 206
      6.5.2   Ternary Systems, 211
      6.5.3   Multicomponent Systems, 219
      6.5.4   Approximate Calculations, 220
      6.5.5   Process Applications, 225

7 Solid–Liquid Equilibrium   228

  7.1 Introduction 228
  7.2 Types of Phase Diagram, 228
  7.3 Experimental Solid–Liquid Equilibrium Data, 233
      7.3.1   Use of Heating and Cooling Curves, 233
      7.3.2   Concentration Cells, 235
      7.3.3   Direct Measurement of Equilibrium Compositions, 236
  7.4 Processing Experimental Data, 236
  7.5 Process Calculations, 237
      7.5.1   Rigorous Calculations, 238
      7.5.2   Approximate Calculations, 253
      7.5.3   Process Applications, 257

8 Special Topics   260

  8.1 Introduction, 260
  8.2 Vapor–Solid Equilibrium, 260
  8.3 Chemical Reaction, 261
      8.3.1   Vapor-Phase Association, 263
      8.3.2   Liquid-Phase Association, 268
      8.3.3   Liquid-Phase Dissociation, 270
  8.4 Conclusion, 270

Index   273

Chapter 1

INTRODUCTION

Equilibrium between mechanically separable phases is a part of the basis for most of the operations involved in chemical engineering. It is not much of an oversimplification to say that chemical engineering consists of the economical application of equations of balance (material, energy, and force), equations of rate (chemical kinetics, heat and mass transfer), and equations of equilibrium (chemical and phase equilibrium).

Most of the separations operations require a favorable relationship between the equilibrium states of two phases for their success. In distillation a difference in composition between a liquid and a gas phase in equilibrium is essential. The same is true of gas absorption. Liquid extraction requires a difference in composition between two liquid phases; crystallization and zone refining involve a liquid and a solid phase. Other examples are possible.

In addition to the separations, some chemical reactors involve more than one phase; the departure from phase equilibrium is just as important in these cases as the kinetics of the reactions within a single phase. The gas-sparged reactor is one of the many examples of such equipment.

Although the equipment in a plant or laboratory rarely operates at a state of equilibrium between the phases, a knowledge of the relationship between the phases at equilibrium is essential to the understanding of the process. This is true whether one adopts the convention of the equilibrium stage, in which case each compartment of equipment is assumed to attain phase equilibrium and an efficiency applied, or the concept of the transfer unit or mass-transfer coefficients, in which case the rate of change everywhere in a piece of equipment is directly proportional to the departure from equilibrium. Neither case can be analyzed without a quantitative knowledge of the compositions the phases would attain if they were allowed to reach equilibrium.

In the past the chemical engineer has usually relied on extensive and expensive laboratory data, the oversimplified *ideal-solution* concept, or the assumption of constant *distribution coefficients* as the basis for his design where phase-equilibrium data were needed. Because of the expense of the laboratory data, the less reliable assumptions have been used more extensively in industrial design. Although the theory of phase equilibrium has not yet progressed to the point where purely theoretical predictions are possible, many strides have been made in that direction in the last decade or two. It is usually possible today to arrive at a quantitative description of the phase equilibria involved in a given practical problem with a minimum of supporting experimental data. This description will be sufficiently accurate for the design engineer to proceed with confidence, expecting the resulting equipment to function to the specifications required without the use of excessive safety factors to overcome the ignorance of the equilibrium equations. Safety factors used for this purpose are extremely dangerous anyway, since no number of extra stages in a distillation column can make it possible to break an azeotrope.

This book is intended to acquaint the practicing chemical engineer with the theoretical and empirical tools required for the practical evaluation of the problems in phase equilibrium encountered in the design of chemical equipment and the development of chemical processes. This task is neither as difficult as many have assumed nor as simple as others believe. One basic assumption involved is that the phase-equilibrium problem can be solved; therefore it is unnecessary to make it a part of the scale-up problem.

In order to understand the principles developed in this book, an elementary understanding of chemical engineering thermodynamics is necessary; however, the reader need not be a thermodynamics expert. Indeed, if he is an expert, he will find much of the presentation oversimplified. A better than average mathematical ability is also required, but it is not necessary for the engineer confronted with a problem in phase equilibrium to be a mathematician. The use of computers in the solution of problems is a distinct advantage. Many of the equations encountered are cumbersome by most standards, and all attempts to linearize them through reasonable assumptions fail. Computer programming is not an essential part of the principles involved, but it must be recognized that computers will be increasingly utilized in the solution of phase-equilibrium problems. To ignore this tool is to handicap oneself unnecessarily. Several short computer programs are included within the text for illustrative purposes. They are written essentially in FORTRAN IV, although there are some variations in the language to adapt to the software for the several computers used in the course of the work (IBM 704, 7044, and 360 and G.E. remote conversational terminals).

The basic thermodynamic equations and correlations are developed in Chapters 2, 3, and 4, with the following chapters devoted to the development of the principles and techniques useful in the design and development of chemical processes. Chapters 5, 6, 7, and 8 contain many worked out numerical examples to illustrate the principles developed in the preceding chapters.

Chapter 2

BASIC THERMODYNAMICS OF PHASE EQUILIBRIUM

2.1 INTRODUCTION

The term "phase equilibrium" implies a condition of equilibrium between two or more phases. By equilibrium we mean no tendency toward change with time; the term "phase" denotes any quantity of matter that is either homogeneous throughout or contains no discontinuity of properties within the space it occupies. The term "homogeneous" is intended to include portions of material that may have been physically separated from one another but are identical in properties and composition; the alternate specification of no discontinuity of properties throughout allows us to consider as a phase a portion of matter that is undergoing a dynamic change with property gradients existing within it but having no boundary across which properties are discontinuous. In certain applications it is also convenient to consider physically separated regions of material, each of which has similar property values and gradients as a single phase.

In studies of phase equilibrium, however, the phase containing gradients is not considered. Wherever gradients exist there is a tendency for change with time; hence there is no equilibrium. On the other hand, there can be two or more phases, each of which is homogenous throughout, with no tendency for any change in properties with time, even though the phases are in intimate physical contact with one another. The latter is the condition that we denote by the term "phase equilibrium."

In a condition of phase equilibrium there are some properties that are drastically different between the phases, and others that must be identical for all phases to prevent a change of properties within individual phases from occurring.

2.2 IMPORTANT THERMODYNAMIC FUNCTIONS

The foregoing discussion outlines the conditions of phase equilibrium qualitatively; a quantitative description involves the evaluation of certain thermodynamic properties and of the mathematical relationships between them. The most basic property of a closed system is the internal energy E, whose absolute value is never known but whose change is given by the first law of thermodynamics as

$$dE = dQ - dW, \qquad (2\text{-}1)$$

where Q and W represent heat added to and work done by the system, respectively. In the case of a reversible process Equation 2-1 can be written

$$dE = T\,dS - P\,dV, \qquad (2\text{-}2)$$

where T represents temperature (absolute scale), P is system pressure, and V is system volume. The symbol S represents entropy, which is defined mathematically by its relationship to the heat Q and temperature T in a reversible process. In an irreversible process Equations 2-1 and 2-2 both hold, but the correspondence of dQ to $T\,dS$ and dW to $P\,dV$ is no longer valid.

Two other properties frequently used in the discussion of phase equilibrium are the enthalpy H and the free energy F. Both these properties are defined mathematically as a combination of the other basic properties. Enthalpy is defined by the equation

$$H = E + PV. \qquad (2\text{-}3)$$

Its change, then, is expressed by

$$\begin{aligned}dH &= dE + d(PV) \\ &= T\,dS - P\,dV + P\,dV + V\,dP, \\ dH &= T\,dS + V\,dP.\end{aligned} \qquad (2\text{-}4)$$

The free energy is defined by the expression

$$F = H - TS. \qquad (2\text{-}5)$$

Equation 2-5, when differentiated, gives

$$\begin{aligned}dF &= -d(TS) + dH \\ &= -T\,dS - S\,dT + T\,dS + V\,dP, \\ dF &= -S\,dT + V\,dP.\end{aligned} \qquad (2\text{-}6)$$

In the development of the mathematical relationships describing phase equilibrium the free energy receives the greatest attention of all the thermodynamic properties defined above.

2.3 THERMODYNAMIC CONDITIONS FOR PHASE EQUILIBRIUM

When two or more phases are at equilibrium in direct contact across an interface, the potentials for change must be zero. This means that there is no tendency for mass or energy to cross the boundary; thus any transport of material or energy from one phase to the other will be a reversible process. It is evident that in order for a condition of equilibrium to exist the temperatures of the phases must be the same; otherwise an irreversible flow of heat will occur. It is also evident that, if the phases are in direct contact with one another, the pressure must be the same in all phases; otherwise the force imbalance would tend to compress one phase and expand the other, with a net irreversible exchange of energy. These two conditions are not sufficient to ensure phase equilibrium, however, and the remaining conditions required are not quite so self-evident. The classic work of Willard Gibbs [1] defined the remaining conditions that ensure equilibrium between phases.

To visualize the additional conditions for equilibrium, consider three separate thermodynamic systems. Systems I and II will represent two phases in equilibrium. System III will have the same temperature and pressure as systems I and II, but it will be initially void (i.e., zero mass). Now, if an infinitesimal quantity of one of the components, dN_i, is transferred from system I to system III, the changes in certain properties are as follows:

$$dH^{III} = \left(\frac{\partial H^I}{\partial N_i}\right)_{T,P,N_{j \neq i}} dN_i,$$

$$dS^{III} = \left(\frac{\partial S^I}{\partial N_i}\right)_{T,P,N_{j \neq i}} dN_i,$$

$$dF^{III} = \left(\frac{\partial F^I}{\partial N_i}\right)_{T,P,N_{j \neq i}} dN_i,$$

$$dE^{III} = \left(\frac{\partial E^I}{\partial N_i}\right)_{T,P,N_{j \neq i}} dN_i.$$

If the properties of system III now are changed by reversible exchanges of energy with its surroundings, excluding systems I and II, until the properties become compatible for a reversible transfer to system II, infinitesimal quantities of heat and work, dQ and dW, will be involved. The changes in properties of system III are then represented by

$$dH^{III} = dQ \quad \text{(constant-pressure process)},$$

$$dS^{III} = \frac{dQ}{T} \quad \text{(reversible constant-temperature process)},$$

$$dF^{III} = dH^{III} - d(TS^{III}) = dQ - dQ = 0,$$

$$dE^{III} = dQ - dW.$$

If the quantity of material, dN_i, is now transferred to system II, the change of properties of system III is given by

$$dH^{III} = -\left(\frac{\partial H^{II}}{\partial N_i}\right)_{T,P,N_{j\neq i}} dN_i,$$

$$dS^{III} = -\left(\frac{\partial S^{II}}{\partial N_i}\right)_{T,P,N_{j\neq i}} dN_i,$$

$$dF^{III} = -\left(\frac{\partial F^{II}}{\partial N_i}\right)_{T,P,N_{j\neq i}} dN_i,$$

$$dE^{III} = -\left(\frac{\partial E^{II}}{\partial N_i}\right)_{T,P,N_{j\neq i}} dN_i.$$

The combination of processes described above represents a reversible transfer of matter from system I to system II, phases in equilibrium. Furthermore, the state of system III is the same at the end of the transfer as at the beginning. Consequently the net change in properties of system III is zero:

$$dH^{III} = 0 = \left(\frac{\partial H^{I}}{\partial N_i}\right)_{T,P,N_{j\neq i}} dN_i + dQ - \left(\frac{\partial H^{II}}{\partial N_i}\right)_{T,P,N_{j\neq i}} dN_i,$$

$$dS^{III} = 0 = \left(\frac{\partial S^{I}}{\partial N_i}\right)_{T,P,N_{j\neq i}} dN_i + \frac{dQ}{T} - \left(\frac{\partial S^{II}}{\partial N_i}\right)_{T,P,N_{j\neq i}} dN_i,$$

$$dF^{III} = 0 = \left(\frac{\partial F^{I}}{\partial N_i}\right)_{T,P,N_{j\neq i}} dN_i - \left(\frac{\partial F^{II}}{\partial N_i}\right)_{T,P,N_{j\neq i}} dN_i,$$

$$dE^{III} = 0 = \left(\frac{\partial E^{I}}{\partial N_i}\right)_{T,P,N_{j\neq i}} dN_i + dQ - dW - \left(\frac{\partial E^{II}}{\partial N_i}\right)_{T,P,N_{j\neq i}} dN_i.$$

From the foregoing equations a relationship between the properties of the two phases in equilibrium can be deduced:

$$\left(\frac{\partial H^{I}}{\partial N_i}\right)_{T,P,N_{j\neq i}} = \left(\frac{\partial H^{II}}{\partial N_i}\right)_{T,P,N_{j\neq i}} - \frac{dQ}{dN_i},$$

$$\left(\frac{\partial S^{I}}{\partial N_i}\right)_{T,P,N_{j\neq i}} = \left(\frac{\partial S^{II}}{\partial N_i}\right)_{T,P,N_{j\neq i}} - \frac{1}{T}\frac{dQ}{dN_i},$$

$$\left(\frac{\partial E^{I}}{\partial N_i}\right)_{T,P,N_{j\neq i}} = \left(\frac{\partial E^{II}}{\partial N_i}\right)_{T,P,N_{j\neq i}} - \frac{d(Q-W)}{dN_i},$$

$$\left(\frac{\partial F^{I}}{\partial N_i}\right)_{T,P,N_{j\neq i}} = \left(\frac{\partial F^{II}}{\partial N_i}\right)_{T,P,N_{j\neq ii}} \tag{2-7}$$

2.4 The Gibbs-Duhem Equations

Only Equation 2-7 defines a property that is identical in both phases. Since dQ and dW are not necessarily equal or zero, the other properties can be different in the two phases. Gibbs gave this property the name "chemical potential" and the special symbol μ. Thus the conditions for phase equilibrium are

$$T^{\text{I}} = T^{\text{II}}, \tag{2-8a}$$

$$P^{\text{I}} = P^{\text{II}}, \tag{2-8b}$$

$$\mu_i^{\text{I}} = \mu_i^{\text{II}} \quad \text{(for all } i\text{)}, \tag{2-8c}$$

where

$$\mu_i \equiv \left(\frac{\partial F}{\partial N_i}\right)_{T,P,N_{j \neq i}} \tag{2-9}$$

2.4 THE GIBBS–DUHEM EQUATIONS

Before further considering the relationships between phases, it is advantageous to develop some of the equations relating properties within a given phase. The equation for the change of free energy for a closed system,

$$dF = -S\,dT + V\,dP,$$

implies that, at constant mass, system free energy is a function of temperature and pressure only. For an open system it also depends on the number of moles of each component. Thus the functional relationship is

$$F = f(T, P, N_1, N_2, \ldots).$$

Stated in differential form, this is

$$dF = \left(\frac{\partial F}{\partial T}\right)_{P,N_j} dT + \left(\frac{\partial F}{\partial P}\right)_{T,N_j} dP + \sum_i \left(\frac{\partial F}{\partial N_i}\right)_{T,P,N_{j \neq i}} dN_i. \tag{2-10}$$

It is evident from an examination of Equation 2-10 that

$$\left(\frac{\partial F}{\partial T}\right)_{P,N_j} = -S \quad \text{and} \quad \left(\frac{\partial F}{\partial P}\right)_{T,N_j} = V, \tag{2-11}$$

giving

$$dF = -S\,dT + V\,dP + \sum_i \mu_i\,dN_i \tag{2-12}$$

as the equation describing the change in free energy in a system of variable mass.

If a system is built up from zero mass at constant temperature and pressure,

$$dF = \sum_i \mu_i \, dN_i.$$

Starting from zero mass, the initial free energy is zero; thus the integral equation becomes

$$\int_0^F dF = F = \sum_i \int_0^{N_i} \mu_i \, dN_i.$$

With the further specification that the composition expressed in terms of mole fractions x_i be constant during the process,

$$dN_i = d(x_i N_t) = x_i \, dN_t$$

and

$$F = \sum_i \int_0^{N_t} \mu_i x_i \, dN_t = \sum_i x_i \int_0^{N_t} \mu_i \, dN_t.$$

It is an experimental fact that in a system of fixed composition the molal free energy F/N_t is independent of the total number of moles in the system. Thus $F/N_t = \sum (x_i/N_t) \int_0^{N_t} \mu_i \, dN_t$ is not a function of N_t. Consequently $\int_0^{N_t} \mu_i \, dN_t \propto N_t$. This can only be true if μ_i is independent of N_t, in which case

$$\frac{F}{N_t} = \sum_i \mu_i x_i, \tag{2-13}$$

or

$$F = \sum_i \mu_i N_i. \tag{2-14}$$

The above equations, shown here for free energy, are generally true as a relationship between total and partial molal properties.

A very useful equation can be derived by differentiating Equation 2-14 and comparing the result with Equation 2-12.

$$dF = \sum_i \mu_i \, dN_i + \sum_i N_i \, d\mu_i = -S \, dT + V \, dP + \sum_i \mu_i \, dN_i,$$

which gives

$$\sum_i N_i \, d\mu_i = -S \, dT + V \, dP, \tag{2-15a}$$

or

$$\sum_i x_i \, d\mu_i = -s \, dT + v \, dP, \tag{2-15b}$$

where $s = S/N_t$ and $v = V/N_t$.

2.5 Mathematical Relationships Between Phases in Equilibrium

If a specification of constant temperature and pressure is applied, the equations become

$$\sum_i N_i \, d\mu_i = 0, \qquad (2\text{-}16a)$$

or

$$\sum_i x_i \, d\mu_i = 0. \qquad (2\text{-}16b)$$

Equation 2-16b is known as the Gibbs–Duhem equation. Equations 2-15 and 2-16 are rigorous and are often used to test experimental data for thermodynamic consistency. Further reference will be made to these equations in subsequent chapters.

2.5 MATHEMATICAL RELATIONSHIPS BETWEEN PHASES IN EQUILIBRIUM

In order to derive further equations relating the properties of phases in equilibrium, attention must be devoted to an evaluation of the quantities μ_i, since, according to Equation 2-8c,

$$\mu_i^{\mathrm{I}} = \mu_i^{\mathrm{II}}.$$

Directing our attention initially to only one phase, the quantity μ_i for any component can be expressed as the sum of two terms:

$$\mu_i = \mu_i^\circ + \Delta\mu_i,$$

where

$$\Delta\mu_i = \mu_i - \mu_i^\circ.$$

The term μ_i° is commonly called the chemical potential of the component i in the *standard state*. In order to evaluate the quantity properly it is necessary to define the standard state. Since there is by no means a "standard" definition of the standard state, there is considerable confusion as a result. The definition of the standard state is generally taken as the state that simplifies the equations to be derived for the specific problem at hand; it is easy to see the difficulties that can arise as a result of this practice. Throughout this book the standard state is always the pure component at the system temperature and total pressure, and in the same physical state as the phase under consideration. Thus the standard state for any component in a gas mixture is the pure component in the gas phase at system temperature and pressure, whereas the standard state for the same component in a liquid mixture is the pure liquid at system temperature and pressure. In the case of two phases at equilibrium each component will have a different standard state in each phase.

This particular definition of standard state makes possible a more straightforward derivation of the equations of phase equilibrium, although there are some instances in which the standard state is a hypothetical one because one or more of the components do not exist as a pure component under the conditions of temperature and pressure of the solution in the same physical state as the mixture. As an example of this situation, methanol and water solutions exist in the liquid state at atmospheric pressure at all temperatures between the boiling points of the two pure components. However, at all temperatures above the boiling point of methanol pure methanol exists in the gaseous state only, at atmospheric pressure. This difficulty is not formidable in most cases, and the use of this particular definition of the standard state has many advantages in visualization of phase phenomena.

The quantity $\Delta \mu_i$ is known as the partial molal free energy of mixing and is generally evaluated in terms of a quantity γ, known as the activity coefficient, defined by the equation

$$\Delta \mu_i = RT \ln (\gamma_i x_i). \tag{2-17}$$

With this definition the chemical potential is expressed by

$$\mu_i = \mu_i^\circ + RT \ln(\gamma_i x_i). \tag{2-18}$$

When Equation 2-8c is applied to two phases in equilibrium,

$$(\mu_i^\circ)^\mathrm{I} + RT \ln(\gamma_i^\mathrm{I} x_i^\mathrm{I}) = (\mu_i^\circ)^\mathrm{II} + RT \ln(\gamma_i^\mathrm{II} x_i^\mathrm{II}), \tag{2-19}$$

which rearranges to give

$$RT \ln \frac{\gamma_i^\mathrm{I} x_i^\mathrm{I}}{\gamma_i^\mathrm{II} x_i^\mathrm{II}} = (\mu_i^\circ)^\mathrm{II} - (\mu_i^\circ)^\mathrm{I}, \tag{2-20}$$

or

$$\frac{\gamma_i^\mathrm{I} x_i^\mathrm{I}}{\gamma_i^\mathrm{II} x_i^\mathrm{II}} = \exp\left(\frac{\Delta \mu_i^\circ}{RT}\right), \tag{2-21}$$

where

$$\Delta \mu_i^\circ = (\mu_i^\circ)^\mathrm{II} - (\mu_i^\circ)^\mathrm{I}. \tag{2-22}$$

Since, by definition, the standard state is the pure component at system temperature and pressure, $\mu_i^\circ = F_i/N_i$ at system temperature and pressure, giving

$$\frac{\gamma_i^\mathrm{I} x_i^\mathrm{I}}{\gamma_i^\mathrm{II} x_i^\mathrm{II}} = \exp\left(\frac{\Delta F_i}{N_i RT}\right). \tag{2-23}$$

2.6 Ideal and Regular Solutions

Thus a quantitative description of phase equilibrium involves primarily the evaluation of the activity coefficients and the difference in free energy experienced by the pure components in a change of state from the physical state of one phase to that of the other at constant temperature and pressure. The evaluation of the standard-state free-energy change is discussed in greater detail in Chapter 3; equations for activity coefficients are discussed in Chapter 4.

2.6 IDEAL AND REGULAR SOLUTIONS

The concept of ideal solution is useful in many calculations. An ideal solution is best defined as a solution in which all activity coefficients are unity (i.e., $\gamma_i = 1$ for all i). This definition, slightly different from the thermodynamic definition usually used in the development of solution thermodynamics, is consistent with, and derivable from, the usual definition so long as the standard state defined in Section 2.5 is used. The results are identical in practical calculations regardless of which definition is used. The concept and the derivation of the equations are simpler if activity coefficients of unity are used as the definition.

If Equation 2-18 is combined with the relationship $F = \Sigma N_i \mu_i$,

$$F = \sum \mu_i^\circ N_i + RT \sum N_i (\ln x_i + \ln \gamma_i). \qquad (2\text{-}24)$$

For an ideal solution

$$F = \sum \mu_i^\circ N_i + RT \sum N_i \ln x_i. \qquad (2\text{-}25)$$

The free energy of an ideal gas mixture can be calculated. By definition, in an ideal gas mixture each component follows the equation

$$p_i V = N_i RT,$$

and the total mixture obeys the equation

$$PV = N_t RT.$$

By dividing the equation for the components by the total equations, we get

$$\frac{p_i}{P} = \frac{N_i}{N_t} = x_i.$$

By adding the equations for all components, it is also evident that $\Sigma p_i = P$, since

$$V \sum p_i = RT \sum N_i = RT N_t = PV.$$

Thus an ideal gas mixture can be considered to consist of separate quantities N_i of each component at system temperature, but at their partial pressures p_i. The properties of the mixture are found simply by adding the properties of the individual components at pressure p_i and temperature T.

The free energy of each individual component can be related to the standard state (pure component at system temperature and pressure) by considering an isothermal expansion from system pressure to component partial pressure:

$$dF_i = -S_i \, dT + V_i \, dP.$$

Throughout the expansion $dT = 0$, but $V_i = N_i RT/P$. Thus

$$\Delta F_i = N_i RT \int_P^{p_i} \frac{dP}{P},$$

or

$$F_i - \mu_i^\circ N_i = N_i RT \ln\left(\frac{p_i}{P}\right) = N_i RT \ln x_i.$$

Thus for the total mixture

$$F = \sum \mu_i^\circ N_i + RT \sum N_i \ln x_i.$$

The above equation is identical with Equation 2-25; therefore an ideal gas mixture is an ideal solution, and any equations applying to ideal solutions can also be applied to an ideal gas mixture. The converse, however, is not true; there are many ideal solutions that are not ideal gases.

Another concept frequently used in solution theory is that of the "regular" solution. In an ideal gas mixture it can be shown that the heat of mixing is zero. Thus $\Delta F_i = \Delta H_i - \Delta(TS_i) = -T \Delta S_i$, since $\Delta H_i = 0$ and the temperature is constant in the process considered. This leads to the equation

$$\Delta S_i = -N_i R \ln x_i, \tag{2-26}$$

or

$$\Delta S = -R \sum N_i \ln x_i \tag{2-27}$$

for an ideal gas. A solution in which the entropy of mixing is given by Equation 2-27 is, by definition, a regular solution. Thus an ideal gas mixture is both an ideal solution and a regular solution; however, a regular solution is not necessarily an ideal solution. The theory of regular solutions is rather thoroughly developed by Hildebrand and Scott [2].

2.7 FUGACITY

The concept of fugacity was first introduced by Lewis [3] in early work on the theory of solution thermodynamics. Although it is possible to omit

2.7 Fugacity

fugacity from a discussion of phase equilibrium and still develop all the pertinent practical equations, the historical significance of the concept has entrenched the term in the language of phase equilibrium.

In order to develop equations for phase equilibrium it is often necessary to evaluate the free energy F of a solution at some pressure P relative to the free energy of the pure components at a pressure P_{ref} at which the mixture behaves as an ideal gas mixture. To make this evaluation the temperature T is held constant. This can be done via the equation

$$F - F_{\text{ref}} = \int_{P_{\text{ref}}}^{P} V\, dP.$$

But for an ideal gas F_{ref} can be evaluated from Equation 2-25 as

$$F_{\text{ref}} = \sum_j N_j (\mu_j^\circ)_{\text{ref}} + RT \sum_j N_j \ln x_j.$$

Thus

$$F = \sum_j N_j (\mu_j^\circ)_{\text{ref}} + RT \sum_j N_j \ln x_j + \int_{P_{\text{ref}}}^{P} V\, dP.$$

The chemical potential is obtained by differentiation as

$$\mu_i = (\mu_i^\circ)_{\text{ref}} + RT \ln x_i + \int_{P_{\text{ref}}}^{P} \bar{v}_i\, dP,$$

where

$$\bar{v}_i = \left(\frac{\partial V}{\partial N_i}\right)_{T,P,N_{j \neq i}}.$$

The above equation is obtained by noting that

$$\sum_j N_j \frac{\partial \ln x_j}{\partial N_i} = 0.$$

The integral in the above equation can be split into two parts:

$$\mu_i = (\mu_i^\circ)_{\text{ref}} + RT \ln x_i + \int_{P_{\text{ref}}}^{P} \left(\bar{v}_i - \frac{RT}{P}\right) dP + \int_{P_{\text{ref}}}^{P} \frac{RT}{P}\, dP,$$

$$\mu_i = (\mu_i^\circ)_{\text{ref}} + RT \ln(Px_i) - RT \ln P_{\text{ref}}$$
$$+ \int_{0}^{P} \left(\bar{v}_i - \frac{RT}{P}\right) dP - \int_{0}^{P_{\text{ref}}} \left(\bar{v}_i - \frac{RT}{P}\right) dP.$$

For an ideal gas $\bar{v}_i = RT/P$; thus

$$\mu_i = (\mu_i^\circ)_{\text{ref}} - RT \ln P_{\text{ref}} + RT \ln(Px_i) + \int_{0}^{P} \left(\bar{v}_i - \frac{RT}{P}\right) dP.$$

Fugacity is introduced through the definition

$$f_i \equiv Px_i \exp\left[\frac{1}{RT}\int_0^P \left(\bar{v}_i - \frac{RT}{P}\right) dP\right]. \tag{2-28}$$

Thus the chemical potential μ_i is related to fugacity as

$$\mu_i = (\mu_i^\circ)_{ref} - RT \ln P_{ref} + RT \ln f_i. \tag{2-29}$$

In the case of two phases at equilibrium we have $\mu_i^I = \mu_i^{II}$, which leads directly to

$$f_i^I = f_i^{II}. \tag{2-30}$$

Furthermore, if f_i° represents the pure component, we have from Equation 2-28

$$f_i^\circ = P \exp\left[\frac{1}{RT}\int_0^P \left(v_i^\circ - \frac{RT}{P}\right) dP\right]. \tag{2-31}$$

From Equations 2-31 and 2-28 we have further

$$f_i = f_i^\circ x_i \exp\left[\frac{1}{RT}\int_0^P (\bar{v}_i - v_i^\circ) dP\right] \tag{2-32}$$

From Equation 2-29, evaluated for a pure component, we have

$$\mu_i^\circ = (\mu_i^\circ)_{ref} - RT \ln P_{ref} + RT \ln f_i^\circ. \tag{2-33}$$

Since $RT \ln \gamma_i x_i = \mu_i - \mu_i^\circ$, it follows that

$$RT \ln \gamma_i x_i = RT \ln \frac{f_i}{f_i^\circ},$$

or

$$f_i = \gamma_i x_i f_i^\circ. \tag{2-34}$$

In the case of an ideal solution $\gamma_i = 1$ and

$$f_i = f_i^\circ x_i. \tag{2-35}$$

Equations 2-30, 2-34, and 2-35 indicate the utility of the fugacity concept. The use of fugacity in place of pressure in the corresponding equation for ideal gases often yields a valid equation. The most important virtue of the fugacity concept to phase equilibrium is that it allows the equations of ideal-solution behavior to be expressed in similar form to the equations of ideal gas mixtures.

The utility of the fugacity concept is discussed further in subsequent chapters.

2.8 SUMMARY

The most important thermodynamic relationships of phase equilibrium developed in this chapter are as follows:

Definition of chemical potential:

$$\mu_i \equiv \left(\frac{\partial F}{\partial N_i}\right)_{T,P,N_{j \neq i}}. \qquad (2\text{-}9)$$

Conditions for equilibrium between phases:

$$T^I = T^{II}, \qquad (2\text{-}8a)$$
$$P^I = P^{II}. \qquad (2\text{-}8b)$$
$$\mu_i^I = \mu_i^{II} \quad (\text{or} \quad f_i^I = f_i^{II}), \qquad (2\text{-}8c, 2\text{-}30)$$

Conditions for thermodynamic consistency within a phase:

$$\sum x_i \, d\mu_i = -s \, dT + v \, dP. \qquad (2\text{-}15b)$$

Definition of activity coefficient:

$$\mu_i - \mu_i^\circ \equiv RT \ln(\gamma_i x_i). \qquad (2\text{-}17)$$

Relationship between phases at equilibrium:

$$\frac{\gamma_i^I x_i^I}{\gamma_i^{II} x_i^{II}} = \exp\left[\frac{(\mu_i^\circ)^{II} - (\mu_i^\circ)^I}{RT}\right]. \qquad (2\text{-}21, 2\text{-}22)$$

Definition of ideal solution:

$$\gamma_i = 1 \quad (\text{for all } i).$$

$$\bar{v}_i = \left(\frac{\partial V}{\partial N}\right)_{T,P,N_{J \neq i}}$$

Definition of regular solution:

$$\Delta S^M = -R \sum N_i \ln x_i. \qquad (2\text{-}27)$$

Definition of fugacity:

$$f_i = P x_i \exp\left[\frac{1}{RT} \int_0^P \left(\bar{v}_i - \frac{RT}{P}\right) dP\right]. \qquad (2\text{-}28)$$

Relationship between fugacity and activity coefficient:

$$f_i = \gamma_i x_i f_i^\circ. \qquad (2\text{-}34)$$

REFERENCES

[1] J. W. Gibbs, *Collected Works*, Vol. I, Longmans, Green, New York, 1931, pp. 55–354.
[2] J. H. Hildebrand and R. L. Scott, *Regular Solutions*, Prentice-Hall, Englewood Cliffs, N.J., 1962.
[3] G. N. Lewis and M. Randall, *Thermodynamics and the Free Energy of Chemical Substances*, McGraw-Hill Book Company, New York, 1923.

BIBLIOGRAPHY

Denbigh, Kenneth, *The Principles of Chemical Equilibrium*, Cambridge University Press, London, 1955.

Dodge, B. F., *Chemical Engineering Thermodynamics*, McGraw-Hill Book Company, New York, 1944.

Gilmont, R., *Thermodynamic Principles for Chemical Engineers*, Prentice-Hall, Englewood Cliffs, N.J., 1959.

Hougen, O. A., K. M. Watson and R. A. Ragatz, *Chemical Process Principles*, Part II, John Wiley & Sons, New York, 1954.

Prigogine, I., and R. Defay, *Chemical Thermodynamics*, translated by D. H. Everett, Longmans, Green, London, 1954.

Smith, J. M., and H. C. Van Ness, *Introduction to Chemical Engineering Thermodynamics*, McGraw-Hill Book Company, New York, 1959.

Van Ness, H. C., *Classical Thermodynamics of Non-Electrolyte Solutions*, Macmillan, New York, 1964.

PROBLEMS

1. Show that the condition $\mu_i^\mathrm{I} = \mu_i^\mathrm{II}$ also applies as a condition for equilibrium for the case in which two solutions are separated by a semipermeable membrane and each solution is confined to a definite space (constant volume).

2. One statement of the definition of ideal solution is that the partial molal volume $\bar{v}_i$ is equal to the pure component's volume v_i° at all pressures up to the solution pressure. Show that this definition leads directly to $\gamma_i = 1$.

3. Show that in an ideal solution the heat of mixing is zero.

Chapter 3

EVALUATION OF STANDARD-STATE FREE-ENERGY DIFFERENCE

3.1 INTRODUCTION

To engineers the practical value of a quantitative description of phase equilibrium lies in the calculation of the composition of phases in equilibrium. The relationship is expressed most concisely in Equation 2-23, which when rearranged relates the mole fraction in one phase to that of the other as

$$x_i^{\text{I}} = \frac{\gamma_i^{\text{II}} x_i^{\text{II}}}{\gamma_i^{\text{I}}} \exp\left[\frac{(\mu_i^\circ)^{\text{II}} - (\mu_i^\circ)^{\text{I}}}{RT}\right]. \tag{3-1}$$

This equation gives the basis for the thermodynamic definition of the distribution coefficient, K_i, which is defined by the equation

$$x_i^{\text{II}} = K_i x_i^{\text{I}}. \tag{3-2}$$

It readily follows that the distribution coefficient is given by

$$K_i = \frac{\gamma_i^{\text{I}}}{\gamma_i^{\text{II}}} \exp\left[\frac{(\mu_i^\circ)^{\text{I}} - (\mu_i^\circ)^{\text{II}}}{RT}\right]. \tag{3-3}$$

The evaluation of the distribution coefficient therefore involves two types of problem:

1. Determination of the activity coefficients γ_i of each component in each phase.
2. Evaluation of the standard-state free-energy difference $\Delta\mu_i^\circ$ for each component between phase II and phase I.

The latter of the two problems is the easier to solve and is the subject of this chapter.

3.2 VAPOR–LIQUID EQUILIBRIUM

According to the definition of standard state developed in Chapter 2, the liquid-phase standard state for each component is evaluated at system pressure π and temperature T in the pure liquid state. Similarly the vapor-phase standard state is evaluated for the pure gas at pressure π and temperature T. Thus to define $\Delta \mu_i^\circ$ it is sufficient to hypothesize any thermodynamic process that transforms a pure liquid at pressure π and temperature T into a pure gas at the same temperature and pressure, and for which the free-energy change can be evaluated.

One such process may be imagined as follows:

Step 1. Change the liquid pressure from π to the vapor pressure p_i^* at constant temperature.

Step 2. Evaporate the liquid reversibly from saturated liquid to saturated vapor at p_i^* and T.

Step 3. Change the gas pressure from p_i^* to π at constant temperature.

The free-energy change in each step can be evaluated by means of the following equation:

$$d\mu_i = -s_i \, dT + v_i \, dP, \tag{3-4}$$

where the lower case letters denote molal quantities. For step 1 at constant temperature

$$\Delta \mu_i = \int_\pi^{p_i^*} (v_i)_L \, dP,$$

which becomes, for an *incompressible* liquid,

$$\Delta \mu_i = -(v_i)_L (\pi - p_i^*). \tag{3-5}$$

For step 2, at constant temperature and constant pressure,

$$\Delta \mu_i = 0.$$

Finally, for step 3

$$\Delta \mu_i = \int_{p_i^*}^\pi (v_i)_G \, dP.$$

If $(v_i)_G$ is expressed in terms of the compressibility Z_i,

$$(v_i)_G = \frac{Z_i RT}{P}$$

3.2 Vapor-Liquid Equilibrium

and

$$\Delta\mu_i = \int_{p_i^*}^{\pi} \frac{Z_i RT}{P} dP.$$

When slightly rearranged, the above equation becomes

$$\Delta\mu_i = \int_{p_i^*}^{\pi} [1 - (1 - Z_i)] \frac{RT}{P} dP$$

$$= RT \ln \frac{\pi}{p_i^*} - \int_{p_i^*}^{\pi} (1 - Z_i) \frac{RT}{P} dP$$

$$= RT \ln \frac{\pi}{p_i^*} - \int_{p_i^*}^{\pi} \left[\frac{RT}{P} - (v_i)_G\right] dP. \quad (3\text{-}6)$$

The total free-energy change for the entire process is obtained by adding the contributions of the individual steps. Thus

$$(\mu_i^\circ)^{II} - (\mu_i^\circ)^{I} = (v_i)_L(p_i^* - \pi) + RT \ln \frac{\pi}{p_i^*} - \int_{p_i^*}^{\pi} \left[\frac{RT}{P} - (v_i)_G\right] dP \quad (3\text{-}7)$$

and

$$\exp\left[\frac{(\mu_i^\circ)^{I} - (\mu_i^\circ)^{II}}{RT}\right] = \frac{p_i^*}{\pi} \exp\left[\frac{(v_i)_L(\pi - p_i^*)}{RT}\right]$$

$$\times \exp\left\{\frac{1}{RT} \int_{p_i^*}^{\pi} \left[\frac{RT}{P} - (v_i)_G\right] dP\right\}. \quad (3\text{-}8)$$

When substituted into Equations 3-2 and 3-3, Equation 3-8 yields

$$\frac{x_i^{II}}{x_i^{I}} = \frac{\gamma_i^{I} p_i^*}{\gamma_i^{II} \pi} \exp\left[\frac{(v_i)_L(\pi - p_i^*)}{RT}\right] \exp\left\{\frac{1}{RT} \int_{p_i^*}^{\pi} \left[\frac{RT}{P} - (v_i)_G\right] dP\right\}. \quad (3\text{-}9)$$

The convention of most literature on vapor–liquid equilibrium is to designate liquid-phase mole fraction by x and vapor-phase mole fraction by y. Applying this convention, we obtain

$$\frac{y_i}{x_i} = \frac{(\gamma_i)_L}{(\gamma_i)_G} \frac{p_i^*}{\pi} \exp\left[\frac{(v_i)_L(\pi - p_i^*)}{RT}\right] \exp\left\{\frac{1}{RT} \int_{p_i^*}^{\pi} \left[\frac{RT}{P} - (v_i)_G\right] dP\right\} \quad (3\text{-}10)$$

If the vapor phase is an ideal gas, $(\gamma_i)_G = 1$ and $(v_i)_G = RT/P$, which gives for this special case

$$\frac{y_i}{x_i} = (\gamma_i)_L \frac{p_i^*}{\pi} \exp\left[\frac{(v_i)_L}{RT}(\pi - p_i^*)\right]. \quad (3\text{-}11)$$

A further simplification is obtained when $(\pi - p_i^*)(v_i)_L$ is small compared with RT:

$$\frac{y_i}{x_i} = (\gamma_i)_L \frac{p_i^*}{\pi}. \tag{3-12}$$

Equation 3-12 is the most widely used equation of vapor–liquid equilibrium, and the liquid-phase subscript is usually omitted, giving the expression

$$y_i = \frac{\gamma_i p_i^* x_i}{\pi}. \tag{3-13}$$

Whenever Equation 3-13 is used, the pertinent assumptions and conventions must be borne in mind, inasmuch as the equation does not apply in situations to which the assumptions are not applicable. These assumptions and conventions bear repeating:

1. Ideal gas phase.
2. Negligible pressure difference between vapor pressure and system pressure.
3. The term γ_i refers to the liquid phase.

These assumptions are met in many instances; hence the abundant usage of Equation 3-13. Whenever there is any question of the applicability of Equation 3-13, Equation 3-10 may be applied if the liquid phase is incompressible.

An alternative expression of Equation 3-10 can be obtained by using the concept of fugacity. By definition

$$f_i^* = p_i^* \exp\left\{\frac{1}{RT} \int_0^{p_i^*} \left[(v_i)_G - \frac{RT}{P}\right] dP\right\}$$

and

$$f_i^\pi = \pi \exp\left\{\frac{1}{RT} \int_0^\pi \left[(v_i)_G - \frac{RT}{P}\right] dP\right\}.$$

Thus

$$\frac{f_i^*}{f_i^\pi} = \frac{p_i^*}{\pi} \exp\left(\frac{1}{RT} \left\{\int_0^{p_i^*} \left[(v_i)_G - \frac{RT}{P}\right] dP - \int_0^\pi \left[(v_i)_G - \frac{RT}{P}\right] dP\right\}\right),$$

which gives

$$\frac{f_i^*}{f_i^\pi} = \frac{p_i^*}{\pi} \exp\left\{\frac{1}{RT} \int_{p_i^*}^\pi \left[\frac{RT}{P} - (v_i)_G\right] dP\right\}. \tag{3-14}$$

Substitution of Equation 3-14, into Equation 3-10 gives

$$\frac{y_i}{x_i} = \frac{(\gamma_i)_L f_i^*}{(\gamma_i)_G f_i^\pi} \exp\left[\frac{(v_i)_L(\pi - p_i^*)}{RT}\right]. \tag{3-15}$$

3.4 Solid-Liquid Equilibrium

The most frequently used special case of Equation 3-15 involves negligible $(\pi - p_i^*)$ and ideal *solution* in the gas phase:

$$\frac{y_i}{x_i} = \frac{\gamma_i f_i^*}{f_i^\pi}. \tag{3-16}$$

The above equations indicate the utility of the fugacity concept in the description of phase equilibrium. It should also be apparent that fugacity, though useful, is by no means essential to the development of quantitative expressions of phase equilibrium.

3.3 VAPOR–SOLID EQUILIBRIUM

Vapor–solid equilibrium is handled in precisely the same manner as vapor–liquid equilibrium. If the solid phase is assigned the role of the liquid phase in Section 3.2, all the equations developed therein are equally applicable to vapor–solid equilibrium.

3.4 SOLID–LIQUID EQUILIBRIUM

The treatment of solid–liquid equilibrium is similar in principle to that of vapor–liquid equilibrium. With subscripts S designating solid phase and L liquid phase, Equation 2-23 becomes

$$\frac{(\gamma_i)_S (x_i)_S}{(\gamma_i)_L (x_i)_L} = \exp\left[\frac{(\mu_i^\circ)_L - (\mu_i^\circ)_S}{RT}\right]. \tag{3-17}$$

A process must be visualized whereby pure solid at system pressure π and temperature T is transformed to pure liquid at π and T, in order to evaluate $(\mu_i^\circ)_L - (\mu_i^\circ)_S$. Since a melting (or freezing) pressure corresponding to a temperature T is not normally known, a process that is somewhat different from the type used for vapor–liquid equilibrium is used:

Step 1. Heat (or cool) the solid component i at constant pressure from temperature T to the melting point T_i^*.
Step 2. Melt the solid at constant temperature T_i^* and pressure π.
Step 3. Cool (or heat) the liquid at constant pressure from T_i^* to system temperature T.

The free-energy change is again evaluated by means of Equation 3-4 for each step. In step 1, at constant pressure,

$$\Delta \mu_i = -\int_T^{T_i^*} (s_i)_S \, dT. \tag{3-18}$$

In step 2, at constant temperature and pressure,

$$\Delta \mu_i = 0. \tag{3-19}$$

And for step 3

$$\Delta \mu_i = -\int_{T_i^*}^{T} (s_i)_L \, dT. \tag{3-20}$$

By adding Equations 3-18, 3-19, and 3-20, we obtain the desired quantity:

$$(\mu_i^\circ)_L - (\mu_i^\circ)_S = \int_{T}^{T_i^*} [(s_i)_L - (s_i)_S] \, dT. \tag{3-21}$$

At this point the problem of evaluating the difference between pure component liquid and solid free energy at T and π has been replaced by one of evaluating an entropy difference at pressure π and at every temperature between T and T_i^*. Since entropy is no more directly measurable than free energy, another process must be hypothesized to evaluate the entropy difference at pressure π and some temperature T', intermediate between T and T_i^*. The most convenient process is similar to that just cited for the free-energy evaluation:

Step 1. Heat (or cool) the solid reversibly at constant pressure from T' to T_i^*. For this process the entropy change is evaluated from

$$ds_i = \frac{dQ_{\text{rev}}}{T'} = \frac{(c_i)_S \, dT'}{T'},$$

or

$$\Delta s_i = \int_{T'}^{T_i^*} \frac{(c_i)_S \, dT'}{T'}, \tag{3-22}$$

where $(c_i)_S$ is the molal heat capacity of the solid at constant pressure.

Step 2. Melt, reversibly, at temperature T_i^* and pressure π, for which

$$\Delta s_i = \frac{Q_{\text{rev}}}{T_i^*} = \frac{H_{Fi}}{T_i^*}, \tag{3-23}$$

where H_{Fi} is the heat of fusion at the melting point T_i^*.

Step 3. Cool (or heat) at constant pressure from T_i^* to T', for which

$$\Delta s_i = \int_{T_i^*}^{T'} \frac{(c_i)_L \, dT}{T'}. \tag{3-24}$$

The sum of Equations 3-22 through 3-24 gives

$$(s_i)_L - (s_i)_S = \frac{H_{Fi}}{T_i^*} - \int_{T'}^{T_i^*} \frac{[(c_i)_L - (c_i)_S] \, dT'}{T'}. \tag{3-25}$$

3.5 Liquid-Liquid Equilibrium

If the heat-capacity difference is constant, the integrations are rather simple. The entropy difference becomes

$$(s_i)_L - (s_i)_S = \frac{H_{Fi}}{T_i^*} - [(c_i)_L - (c_i)_S] \ln \frac{T_i^*}{T'}. \tag{3-26}$$

Substitution into Equation 3-21 yields

$$(\mu_i^\circ)_L - (\mu_i^\circ)_S = \int_T^{T_i^*} \left\{ \frac{H_{Fi}}{T_i^*} - [(c_i)_L - (c_i)_S] \ln \frac{T_i^*}{T'} \right\} dT'. \tag{3-27}$$

Integration of Equation 3-27 in turn yields

$$(\mu_i^\circ)_L - (\mu_i^\circ)_S = H_{Fi}\left(\frac{T_i^* - T}{T_i^*}\right) + [(c_i)_L - (c_i)_S]\left(T - T_i^* - T \ln \frac{T}{T_i^*}\right). \tag{3-28}$$

Equation 3-28, substituted into Equation 3-17, yields

$$\frac{(\gamma_i)_S(x_i)_S}{(\gamma_i)_L(x_i)_L} = \exp\left(\frac{1}{RT}\left\{\frac{H_{Fi}}{T_i^*} - [(c_i)_L - (c_i)_S]\right\}(T_i^* - T) - \frac{(c_i)_L - (c_i)_S}{R} \ln \frac{T}{T_i^*}\right), \tag{3-29}$$

which also gives

$$\frac{(\gamma_i)_S(x_i)_S}{(\gamma_i)_L(x_i)_L} = \left(\frac{T}{T_i^*}\right)^{-[(c_i)_L - (c_i)_S]/R} \exp\left\{\left(\frac{T_i^* - T}{RT}\right)\left[\frac{H_{Fi}}{T_i^*} - (c_i)_L + (c_i)_S\right]\right\}. \tag{3-30}$$

Equation 3-30 is applicable only to cases for which $(c_i)_L - (c_i)_S$ is independent of temperature. The treatment of cases in which this is not true is just as straightforward, although the final expression is somewhat more complex. The procedure is to substitute the appropriate expressions for heat capacity into Equation 3-25, integrate, substitute the result into Equation 3-21, and integrate again.

3.5 LIQUID–LIQUID EQUILIBRIUM

For equilibrium between two liquid phases the standard state is the same for both phases, giving

$$(\mu_i^\circ)^\mathrm{I} - (\mu_i^\circ)^\mathrm{II} = 0.$$

This simple result yields the equally simple equation

$$\gamma_i^\mathrm{I} x_i^\mathrm{I} = \gamma_i^\mathrm{II} x_i^\mathrm{II}. \tag{3-31}$$

3.6 SOLID–SOLID EQUILIBRIUM

The preceding discussion also applies to two solid phases if the same crystalline configuration is used as the standard state for both phases. Thus Equation 3-31 applies. If a different configuration is used for the standard state, an equation similar to Equation 3-30 can be used where T_i^* is replaced by the temperature at which the solid-state transition between the two crystalline states occurs and the liquid heat capacity is replaced by the heat capacity of the second crystalline state.

3.7 GAS–LIQUID EQUILIBRIUM

It has been already stated that in certain cases the standard state, as we have defined it, does not actually exist. In most cases the value μ_i° can nevertheless be evaluated by a small extrapolation of the equation of state. In Equation 3-10, for example, if π is greater than p_i^*, the gas phase of pure component i does not actually exist at the pressure π. However, an extrapolation of $(v_i)_G$ to higher pressures can often be made without much sacrifice in accuracy. There are cases, however, where this extrapolation cannot be made; one such example is equilibrium between a gas well above its critical temperature and a nonvolatile liquid. Suppose we attempt to evaluate $(\mu_i^\circ)_G - (\mu_i^\circ)_L$ in this case, where component i is the gas in question. Since the liquid phase for component i does not exist at any pressure at system temperature, $(v_i)_L$ and p_i^* are not defined. There is, in fact, no way to determine $(\mu_i^\circ)_L$ or $(\mu_i^\circ)_G - (\mu_i^\circ)_L$, yet we must have this value if we are to use the equation

$$\frac{(\gamma_i)_L x_i}{(\gamma_i)_G y_i} = \exp\left[\frac{(\mu_i^\circ)_G - (\mu_i^\circ)_L}{RT}\right]. \tag{3-32}$$

The dilemma can be averted if we define a state point P as the saturated solution of i in the solvent at a pressure π_P sufficiently low for Henry's law to apply. Thus

$$\pi y_i = H_i x_i \quad \text{for } \pi \leq \pi_P.$$

The standard-state free-energy difference can be expressed as the sum of two parts:

$$(\mu_i^\circ)_G - (\mu_i^\circ)_L = (\mu_i^\circ)_G - (\mu_i)_P + (\mu_i)_P - (\mu_i^\circ)_L.$$

3.7 Gas-Liquid Equilibrium

But

$$(\mu_i)_P = (\mu_i^\circ)_L + RT \ln[(\gamma_i)_P(x_i)_P] + \int_\pi^{\pi_P} (\bar{v}_i)_L \, dP,$$

where the last term corrects the chemical potential from system pressure to π_P. If $(\bar{v}_i)_L$ is assumed to have a constant average value,

$$(\mu_i)_P - (\mu_i^\circ)_L = RT \ln[(\gamma_i)_P(x_i)_P] - (\bar{v}_i)_L(\pi - \pi_P).$$

Since the state point P is one at which the liquid is at equilibrium with a pure gas at pressure π_P, $(\mu_i)_P$ is also the chemical potential of the pure gaseous component i at the pressure π_P. Thus

$$(\mu_i^\circ)_G - (\mu_i)_P = \int_{\pi_P}^\pi (v_i)_G \, dP$$

and

$$(\mu_i^\circ)_G - (\mu_i^\circ)_L = \int_{\pi_P}^\pi (v_i)_G \, dP + RT \ln[(\gamma_i)_P(x_i)_P] - (\bar{v}_i)_L(\pi - \pi_P).$$

By substitution into Equation 3-32, we have

$$\frac{(\gamma_i)_L x_i}{(\gamma_i)_G y_i} = (\gamma_i)_P(x_i)_P \exp\left\{\frac{1}{RT}\left[\int_{\pi_P}^\pi (v_i)_G \, dP - (\bar{v}_i)_L(\pi - \pi_P)\right]\right\}.$$

Applying Henry's law, we obtain

$$(x_i)_P = \frac{\pi_P}{H_i},$$

giving

$$\frac{(\gamma_i)_L x_i}{(\gamma_i)_G y_i} = \frac{(\gamma_i)_P \pi_P}{H_i} \exp\left\{\frac{1}{RT}\left[\int_{\pi_P}^\pi (v_i)_G \, dP - (\bar{v}_i)_L(\pi - \pi_P)\right]\right\}.$$

Furthermore, if π_P is chosen sufficiently low for the solution to approach infinite dilution, $(\gamma_i)_P \to (\gamma_i^\infty)_L$, where γ_i^∞ is defined as the activity coefficient at infinite dilution. If we further assume an ideal solution in the gas phase and drop the L subscript, we have

$$\frac{\gamma_i x_i}{y_i} = \frac{\gamma_i^\infty \pi_P}{H_i} \exp\left\{\frac{1}{RT}\left[\int_{\pi_P}^\pi (v_i)_G \, dP - (\bar{v}_i)_L(\pi - \pi_P)\right]\right\}.$$

If we further assume that $\pi - \pi_p$ is small compared with RT,

$$\frac{\gamma_i x_i}{y_i} = \frac{\gamma_i^\infty \pi_P}{H_i} \exp\left[\frac{1}{RT}\int_{\pi_P}^\pi (v_i)_G \, dP\right]. \tag{3-33}$$

And, if the gas phase is ideal, this becomes

$$\frac{\gamma_i x_i}{y_i} = \frac{\gamma_i^\infty}{H_i} \pi,$$

or

$$x_i = \frac{\gamma_i^\infty \pi y_i}{H_i \gamma_i}.$$

In terms of fugacity, Equation 3.33 is

$$\frac{\gamma_i x_i}{y_i} = \frac{\gamma_i^\infty \pi_P}{H_i} \frac{f_i^\pi}{f_i^P}.$$

Since it is generally possible to choose π_P sufficiently small for $f_i^P = \pi_P$,

$$x_i = \frac{\gamma_i^\infty}{\gamma_i} \frac{f_i^\pi}{H_i} y_i. \tag{3-34}$$

In Equation 3-34 we see that the quantity H_i/γ_i^∞ corresponds to the extrapolated vapor pressure p_i^*. Hence $\gamma_i^\infty = H_i/p_i^*$. By correlating data in terms of p_i^* and γ_i^∞, however, we can maintain the same standard state for component i, regardless of the solvent in which it is dissolved. With the frequently used alternative standard state of infinite dilution for the dissolved gas, the standard state differs with the solvent, and multicomponent solvents present a dilemma.

3.8 SUMMARY

The equations describing phase equilibrium have been evaluated for the standard-state free-energy difference. The resulting equations of particular importance are the following:

General equation for vapor–liquid equilibrium:

$$\frac{y_i}{x_i} = \frac{(\gamma_i)_L}{(\gamma_i)_G} \frac{p_i^*}{\pi} \exp\left[\frac{(v_i)_L(\pi - p_i^*)}{RT}\right] \exp\left\{\frac{1}{RT}\int_{p_i^*}^{\pi}\left[\frac{RT}{P} - (v_i)_G\right] dP\right\}. \tag{3-10}$$

Vapor–liquid equilibrium, ideal gas phase and negligible difference between system pressure and vapor pressure:

$$y_i = \frac{\gamma_i p_i^* x_i}{\pi}. \tag{3-13}$$

Vapor–liquid equilibrium, ideal solution in gas phase and neglecting pressure difference, expressed in terms of fugacities:

3.8 Summary

$$\frac{y_i}{x_i} = \frac{\gamma_i f_i^*}{f_i^\pi}. \qquad (3\text{-}16)$$

Solid–liquid equilibrium with constant heat-capacity difference:

$$\frac{(\gamma_i)_S(x_i)_S}{(\gamma_i)_L(x_i)_L} = \left(\frac{T}{T_i^*}\right)^{-[(c_i)_L - (c_i)_S]/R} \exp\left\{\left(\frac{T_i^* - T}{RT}\right)\left[\frac{H_{Fi}}{T_i^*} - (c_i)_L + (c_i)_S\right]\right\}. \qquad (3\text{-}30)$$

Liquid–liquid or solid–solid equilibrium:

$$\gamma_i^{\,\text{I}} x_i^{\,\text{I}} = \gamma_i^{\,\text{II}} x_i^{\,\text{II}}. \qquad (3\text{-}31)$$

For vapor–solid equilibrium the equations of vapor–liquid equilibrium also apply.

For gas solubility of permanent gases:

$$x_i = \frac{\gamma_i^\infty}{\gamma_i} \frac{f_i^\pi}{H_i} y_i. \qquad (3\text{-}34)$$

PROBLEMS

1. Derive an expression for standard-state free-energy difference for vapor–liquid equilibrium in terms of the temperature at which the vapor pressure is equal to the system pressure.

2. Derive an expression for standard-state free-energy difference for solid–liquid equilibrium with a solid-state transition existing between the system temperature and the melting temperature.

3. Derive an expression for standard-state free-energy difference between two solid phases for the case in which the crystalline configuration for the standard state is not the same in both phases.

4. Derive the expression for solid–liquid standard-state free-energy change when the heat capacities are linear functions of temperature.

5. Assuming that benzene and toluene form an ideal solution, calculate the vapor composition in equilibrium with liquid having mole fractions of 0.1, 0.3, 0.5, and 0.8 benzene at 100°C. Also calculate the total pressure over each solution.

6. Assume that pure ice freezes out of an aqueous solution, which forms an ideal liquid-phase solution. Calculate the composition of the liquid phase in equilibrium with the ice at -10, -20, and $-30°C$.

7. Assume an ideal solution in the liquid phase and ideal gas phase in equilibrium. Assume component i to have a vapor pressure of 1 atmosphere at 25°C. Calculate and plot the ratio y_i/x_i as a function of total pressure up to 1000 atmospheres. The liquid is assumed to be incompressible.

Chapter 4

EQUATIONS FOR ACTIVITY COEFFICIENTS

4.1 INTRODUCTION

The evaluation of standard-state free-energy differences, as outlined in Chapter 3, leaves only the problem of evaluating the activity coefficient of each component in each phase before a complete solution of Equation 3-3 is attained. This is, by far, the most difficult problem of phase equilibrium, and its solution is by no means complete at this time.

The utopian situation would exist if we could, from a knowledge of the pure component's thermodynamic properties and composition of a mixture, calculate the activity coefficients without recourse to any further data. Some definite progress is being made toward that end, but to allow successful design of process equipment, some experimental data are still necessary in most cases.

This chapter discusses the methods available for treating solutions of each of the three states of matter (solid, liquid, and gas). There is a lot of research activity in progress, particularly for the liquid state; therefore it is quite possible that some of the methods in current use will be replaced by better methods in the near future. However, the discussion here deals with the methods and equations of greatest practical utility at the time of this writing.

4.2 GAS-PHASE SOLUTIONS

In order to determine the activity coefficient of a gas mixture we must start with the definition of the activity coefficient, given by Equation 2-17, slightly rearranged:

$$\ln \gamma_i = \frac{\mu_i - \mu_i^\circ}{RT} - \ln x_i. \tag{4-1}$$

4.2 Gas-phase Solutions

The activity coefficient can be evaluated if we can determine the difference between μ_i, the chemical potential of component i in the mixture, and μ_i°, the molal free energy of pure component i at system temperature and pressure. To evaluate this difference we can make use of the experimental observation that, as zero pressure is approached, the behavior of real gases approaches that of ideal gases. Thus, if we take as a reference the pure gases of the mixture at zero pressure and system temperature T, we have a basis for evaluating the chemical potentials. If N_i moles of component i are compressed from zero pressure to system pressure π, the free-energy change is

$$\Delta F_i = N_i(\mu_i^\circ - [\mu_i^\circ]_{p=0}) = N_i \int_0^\pi v_i \, dP. \tag{4-2}$$

Rearrangement yields

$$\mu_i^\circ = [\mu_i^\circ]_{p=0} + \int_0^\pi v_i \, dP. \tag{4-3}$$

Again starting at zero pressure, we can mix the N_i moles of each component to form an ideal gas mixture. Since an ideal gas mixture is an ideal solution, Equation 2-25 applies:

$$F = \sum [\mu_i^\circ]_{p=0} N_i + RT \sum N_i \ln x_i. \tag{2-25}$$

If the gas mixture is then compressed to pressure π, the free-energy change is

$$\Delta F = N_t \int_0^\pi v_M \, dP, \tag{4-4}$$

where v_M is the molal volume of the gas mixture. Addition of Equations 2-25 and 4-4 results in

$$F = \sum N_i [\mu_i^\circ]_{p=0} + RT \sum N_i \ln x_i + N_t \int_0^\pi v_M \, dP. \tag{4-5}$$

The definition of μ_i is

$$\mu_i = \left(\frac{\partial F}{\partial N_i}\right)_{T,P,N_j \neq i}.$$

Thus μ_i is evaluated by differentiating Equation 4-5:

$$\mu_i = [\mu_i^\circ]_{p=0} + RT \ln x_i + \int_0^\pi \bar{v}_i \, dP, \tag{4-6}$$

where

$$\bar{v}_i \equiv \left(\frac{\partial (V_M)}{\partial N_i}\right)_{T,P,N_j \neq i}.$$

The term V_M represents the total volume of the mixture and is identical to $N_t v_M$. Equation 4-3 is subtracted from Equation 4-6 to yield

$$\mu_i - \mu_i^\circ = RT \ln x_i + \int_0^\pi (\bar{v}_i - v_i) \, dP. \tag{4-7}$$

Equation 4-7 is substituted into Equation 4-1 to produce

$$\ln \gamma_i = \frac{1}{RT} \int_0^\pi (\bar{v}_i - v_i) \, dP. \tag{4-8}$$

The molal volume v_i of the pure component is known if the equation of state for the pure component is known. The partial molal volume $\bar{v}_i$ can be obtained if an equation of state for the mixture is known, but an additional equation must first be derived. By definition we have

$$\bar{v}_i = \frac{\partial (N_t v_M)}{\partial N_i},$$

which can be manipulated as follows:

$$\bar{v}_i = N_t \frac{\partial v_M}{\partial N_i} + v_M \frac{\partial N_t}{\partial N_i},$$

$$N_t = N_1 + N_2 + \cdots + N_i + \cdots,$$

$$\frac{\partial N_t}{\partial N_i} = 1.$$

Therefore

$$\bar{v}_i = v_M + N_t \frac{\partial v_M}{\partial N_i} = v_M + N_t \sum_j \frac{\partial v_M}{\partial x_j} \frac{\partial x_j}{\partial N_i}, \tag{4-9}$$

$$x_j = \frac{N_j}{N_t},$$

$$\frac{\partial x_j}{\partial N_i} = -\frac{N_j}{(N_t)^2} = -\frac{x_j}{N_t} \quad \text{(when } j \neq i\text{)},$$

$$\frac{\partial x_i}{\partial N_i} = \frac{N_t - N_i}{(N_t)^2} = \frac{1}{N_t}(1 - x_i) \quad \text{(when } j = i\text{)}.$$

Substitution of the above values of $\partial x_j / \partial N_j$ into Equation 4-9 yields

$$\bar{v}_i = v_M - x_1 \frac{\partial v_M}{\partial x_1} - x_2 \frac{\partial v_M}{\partial x_2} + \cdots + \frac{\partial v_M}{\partial x_i}(1 - x_i) + \cdots,$$

or

$$\bar{v}_i = v_M + \frac{\partial v_M}{\partial x_i} - \sum_j x_j \frac{\partial v_M}{\partial x_j}. \tag{4-10}$$

4.2 Gas-Phase Solutions

Equation 4-10 is a general equation, applicable to all partial molal properties. Application can be made to properties other than volume by merely substituting the appropriate property for volume in Equation 4-10. Substitution of Equation 4-10 into Equation 4-8 yields

$$RT \ln \gamma_i = \int_0^\pi (v_M - v_i)\, dP + \int_0^\pi \frac{\partial v_M}{\partial x_i}\, dP - \sum_j x_j \int_0^\pi \frac{\partial v_M}{\partial x_j}\, dP. \quad (4\text{-}11)$$

Thus an equation of state for the mixture giving v_M as a function of pressure and composition (the x_j terms) at the system temperature T is sufficient for the determination of the gas-phase activity coefficient.

If a volume-explicit equation of state is available, the integration of Equation 4-11 is relatively straightforward. A frequently used equation of state is the virial equation truncated after the first-order term:

$$\frac{Pv}{RT} = 1 + \frac{B}{v}. \quad (4\text{-}12)$$

If B is small relative to v, an approximate solution, explicit in v, is

$$v = \frac{RT}{P} + B. \quad (4\text{-}13)$$

Substitution into Equation 4-11 gives

$$RT \ln \gamma_i = \int_0^\pi (B_M - B_{ii})\, dP + \int_0^\pi \frac{\partial B_M}{\partial x_i}\, dP - \sum_j x_j \int_0^\pi \frac{\partial B_M}{\partial x_j}\, dP, \quad (4\text{-}14)$$

where B_{ii} represents the second virial coefficient of pure component i and B_M that of the mixture. The coefficient B_M, according to the statistical mechanics derivation of the virial equation, is given by

$$B_M = \sum_k \sum_j x_k x_j B_{kj}, \quad (4\text{-}15)$$

with $B_{kj} = B_{jk}$.

If Equation 4-14 is written as a single integral, it becomes

$$RT \ln \gamma_i = \int_0^\pi \Bigg(-B_{ii} + \sum_k \sum_j x_k x_j B_{kj} + \frac{\partial}{\partial x_i} \sum_k \sum_j x_k x_j B_{kj}$$

$$- \sum_m x_m \frac{\partial}{\partial x_m} \sum_k \sum_j x_k x_j B_{kj} \Bigg) dP. \quad (4\text{-}16)$$

Differentiation gives

$$\frac{\partial}{\partial x_m} \sum_j \sum_k x_k x_j B_{kj} = 2 \sum_k x_k B_{km}. \quad (4\text{-}17)$$

Equation 4-17 transforms Equation 4-16 to the following form:

$$RT \ln \gamma_i = \int_0^\pi \left(-B_{ii} - \sum_k \sum_j x_k x_j B_{kj} + 2 \sum_k x_k B_{ik} \right) dP. \quad (4\text{-}18)$$

Since the B and x terms are not functions of pressure in this integration, the activity coefficient of the gas mixture is given by

$$\ln \gamma_i = \frac{\pi}{RT} \left(2 \sum_k x_k B_{ik} - B_{ii} - \sum_k \sum_j x_k x_j B_{kj} \right). \quad (4\text{-}19)$$

For a binary system we may write

$$\ln \gamma_1 = (2x_1 B_{11} + 2x_2 B_{12} - B_{11} - x_1{}^2 B_{11} - 2x_1 x_2 B_{12} - x_2{}^2 B_{22}) \frac{\pi}{RT}$$

$$= [2B_{12}(x_2 - x_1 x_2) - B_{11}(x_1{}^2 - 2x_1 + 1) - B_{22} x_2{}^2] \frac{\pi}{RT},$$

or

$$\ln \gamma_1 = (2B_{12} - B_{11} - B_{22}) \frac{\pi}{RT} x_2{}^2. \quad (4\text{-}20)$$

It is usually assumed that $B_{12} = \sqrt{B_{11} B_{22}}$, which gives

$$\ln \gamma_1 = \frac{\pi}{RT} (-B_{11} + 2\sqrt{B_{11} B_{22}} - B_{22}) x_2{}^2. \quad (4\text{-}21)$$

Equation 4-21 can be evaluated entirely from the virial coefficients of the pure components, requiring no experimental determination. Its applicability is dependent on the applicability of Equation 4-13 and of the geometric mean assumption for the virial coefficients of the mixture. These conditions are generally satisfied only at low pressures.

When a pressure-explicit equation of state must be used, direct integration of Equation 4-11 is cumbersome in that a numerical solution for v_M and v_i must be made at every increment for a numerical integration. It is possible, however, to replace the iterative numerical solutions to the equation of state by a set of solutions at the final pressure only. The procedure will be developed for a general two-parameter equation of state with a specific application to the Redlich–Kwong [1] equation. The extension of the technique to other equations of state or to three or more parameters is straightforward.

The type of equation of state to be considered can be written, in general, as

$$P = f(v, a, b). \quad (4\text{-}22)$$

In Equation 4-11 pressure P is an independent variable and v is a function of P and composition, designated by

$$v = v(P, x).$$

4.2 Gas-Phase Solutions

The parameters a and b are independent of pressure but are functions of composition:

$$a = a(x); \qquad b = b(x).$$

The use of x in the functional solution is intended to designate *all* of the individual mole fractions x_j. In the derivation to follow the notation f_i is used to designate the function $f(v, a, b)$ when pure-component parameters a_i and b_i are used; the shorthand symbol f designates the use of mixture parameters a_M and b_M.

The first two terms of Equation 4-11 can be integrated by parts to give

$$\int_0^\pi v_M \, dP - \int_0^\pi v_i \, dP = \pi(v_M - v_i) + \int_{v_M}^\infty f \, dv + \int_\infty^{v_i} f_i \, dv. \qquad (4\text{-}23)$$

Equation 4-23 can be integrated directly once the form of the function f is known. The integrations indicated are to be done at constant composition and constant temperature.

The remaining terms of Equation 4-11 all involve integration of terms of the type

$$\int_0^\pi \frac{\partial v}{\partial x_i} \, dP.$$

Differentiation of Equation 4-22 yields

$$dP = \frac{\partial f}{\partial v} \, dv + \frac{\partial f}{\partial a} \, da + \frac{\partial f}{\partial b} \, db. \qquad (4\text{-}24)$$

For use in the integration, to be done at constant composition, da and db are zero and

$$dP = \frac{\partial f}{\partial v} \, dv. \qquad (4\text{-}25)$$

On the other hand, if the differentiation of Equation 4-24 is with respect to one of the mole fractions x_i, dP is zero and

$$\frac{\partial v}{\partial x_i} = -\frac{(\partial f/\partial a)(\partial a/\partial x_i) + (\partial f/\partial b)(\partial b/\partial x_i)}{(\partial f/\partial v)}. \qquad (4\text{-}26)$$

Application of Equation 4-25 yields

$$\int_0^\pi \frac{\partial v_i}{\partial x_i} \, dP = \frac{\partial a}{\partial x_i} \int_{v_M}^\infty \frac{\partial f}{\partial a} \, dv + \frac{\partial b}{\partial x_i} \int_{v_M}^\infty \frac{\partial f}{\partial b} \, dv. \qquad (4\text{-}27)$$

Substitution of Equations 4-27 and 4-23 into Equation 4-11 gives the activity coefficient:

$$RT \ln \gamma_i = \pi(v_M - v_i) + \int_{v_M}^{\infty} f \, dv + \int_{\infty}^{v_i} f_i \, dv$$

$$+ \left(\frac{\partial a}{\partial x_i} - \sum_j x_j \frac{\partial a}{\partial x_j}\right) \int_{v_M}^{\infty} \frac{\partial f}{\partial a} \, dv + \left(\frac{\partial b}{\partial x_i} - \sum_j x_j \frac{\partial b}{\partial x_j}\right) \int_{v_M}^{\infty} \frac{\partial f}{\partial b} \, dv. \quad (4\text{-}28)$$

Further development of Equation 4-28 requires a "mixing rule" to give the dependence of the parameters a and b on the composition. Equation 4-15 is a mixing rule for the virial equation of state. There is no general mixing rule that is applicable to all equations of state, whose determination is a matter for experimental verification. In the absence of a general rule a simple linear combination of pure-component parameters is used:

$$a_M = \sum_j x_j a_j; \quad \frac{\partial a}{\partial x_k} = a_k,$$

$$b_M = \sum_j x_j b_j; \quad \frac{\partial b}{\partial x_k} = b_k. \quad (4\text{-}29)$$

With the above equations, it is evident that

$$\sum_j x_j \frac{\partial a}{\partial x_j} = \sum_j x_j a_j = a_M$$

and

$$\sum_j x_j \frac{\partial b}{\partial x_j} = b_M. \quad (4\text{-}30)$$

Equation 4-30 gives the following simplification of Equation 4-28:

$$RT \ln \gamma_i = \pi(v_M - v_i) + \int_{v_M}^{\infty} f \, dv + \int_{\infty}^{v_i} f_i \, dv$$

$$+ (a - a_M) \int_{v_M}^{\infty} \frac{\partial f}{\partial a} \, dv + (b - b_M) \int_{v_M}^{\infty} \frac{\partial f}{\partial b} \, dv. \quad (4\text{-}31)$$

Equation 4-31 cannot be significantly simplified without the application of the specific equation of state to be used. The Redlich–Kwong equation of state can be written as

$$P = f(v, a, b) = \frac{RT}{v - b} + \frac{a}{\sqrt{T} \, v(v + b)}$$

$$= \frac{RT}{v - b} + \frac{a}{\sqrt{T} \, b} \left(\frac{1}{v} - \frac{1}{v + b}\right). \quad (4\text{-}32)$$

4.2 Gas-Phase Solutions

By integrating Equation 4-32 and applying the appropriate limits as infinite volume is approached, we can evaluate the first two terms of Equation 4-31:

$$\int_{v_M}^{\infty} f \, dv + \int_{\infty}^{v_i} f_i \, dv = RT \ln \frac{v_i - b_i}{v_M - b_M} + \frac{1}{\sqrt{T}} \frac{a_i}{b_i} \ln \left(\frac{v_i}{v_i + b_i} - \frac{a_M}{b_M} \ln \frac{v_M}{v_M + b_M} \right). \tag{4-33}$$

The third term of Equation 4-31 requires the partial derivative with respect to the parameter a:

$$\frac{\partial f}{\partial a} = \frac{1}{b_M \sqrt{T}} \left(\frac{1}{v} - \frac{1}{v + b_M} \right). \tag{4-34}$$

The integral of this term thus becomes

$$\int_{v_M}^{\infty} \frac{\partial f}{\partial a} \, dv = -\frac{1}{b_M \sqrt{T}} \ln \frac{v_M}{v_M + b_M}. \tag{4-35}$$

Similarly the integral of the fourth term is evaluated:

$$\frac{\partial f}{\partial b} = \frac{RT}{(v-b)^2} - \frac{a}{\sqrt{T} v(v+b)^2} \tag{4-36}$$

Thus

$$\int_{v_M}^{\infty} \frac{\partial f}{\partial b} \, dv = \frac{RT}{v_M - b_M} + \frac{a_M}{\sqrt{T} v_M (v_M + b_M)}. \tag{4-37}$$

From Equation 4-32 we can derive

$$\frac{a_M}{\sqrt{T} v_M (v_M + b_M)} = \left(\pi - \frac{RT}{v_M - b_M} \right) \frac{v_M}{b_M}. \tag{4-38}$$

By substitution of Equation 4-38 into Equation 4-37, we obtain

$$\int_{v_M}^{\infty} \frac{\partial f}{\partial b} \, dv = \frac{\pi v_M - RT}{b_M} + \frac{a_M}{b_M^2 \sqrt{T}} \ln \frac{v_M}{v_M + b_M}. \tag{4-39}$$

Equations 4-39, 4-35, and 4-33 are substituted into Equation 4-31 to yield, after some rearrangement,

$$\ln \gamma_i = \frac{\pi}{RT b_M} (b_i v_M - b_M v_i) + \frac{b_M - b_i}{b_M} + \ln \frac{v_i - b_i}{v_M - b_M}$$

$$+ \frac{1}{RT^{1.5}} \left[\frac{a_i}{b_i} \ln \frac{v_i}{v_i + b_i} + \frac{(b_i + b_M) a_M - a_i b_M}{b_M^2} \ln \frac{v_M}{v_M + b_M} \right]. \tag{4-40}$$

Throughout all of the derivations above it has been assumed that in the limit of infinite volume, or zero pressure, the ideal-gas law applies. This is a characteristic of all successful equations of state for gas mixtures and introduces no limitation on the applicability of the equations. The procedure used for evaluating activity coefficients from pressure-explicit equations of state can be summarized as follows:

1. Change the variable of integration from pressure to volume. This results in the generally applicable Equation 4-28, which can be extended to more than three parameters by including more terms of the form of the final two.

2. Formulate a mixing rule for the evaluation of the dependence of the parameters on composition. For the specific mixing rule used herein Equation 4-31 results.

3. Apply the specific equation of state to determine the parametric partial derivatives and perform the indicated integration at constant temperature and constant composition. Make use of the ideal-gas law at infinite volume. This step results in Equation 4-40 for the Redlich–Kwong equation of state.

The current tendency in the literature is to use the virial equation of state, truncated after the second or third term until a reduced pressure of 0.5 is reached. The virial equation is properly expressed in a pressure-explicit form, although a volume-explicit approximation is sometimes used. Better results are obtained from the pressure-explicit form, for which theoretically derived mixing rules are available.

At reduced pressures up to 0.8 the Redlich–Kwong equation, with parameters evaluated from the critical constants of the pure components, is often used. The mixing rules are empirical and still the subject of experimental investigation. The current trend is toward a quadratic representation for the Redlich–Kwong a:

$$a_M = \sum_i \sum_j a_{ij} y_i y_j,$$

where $a_{ij} = (1 - k_{ij})\sqrt{a_{ii} a_{jj}}$ and a_{ii} is the pure-component parameter. The k_{ij} is a small, empirically determined parameter [2].

Another approach often used in gas-mixture calculations to circumvent the problem of evaluating the standard-state chemical potential μ_i° is to take advantage of Equation 2-34, expanded as follows:

$$f_i = \gamma_i x_i f_i^\circ$$

$$= \gamma_i x_i \left(\frac{f_i^\circ}{\pi}\right) \pi$$

$$= \phi_i \pi x_i$$

where

$$\phi_i \equiv \gamma_i \left(\frac{f_i^\circ}{\pi}\right).$$

By manipulation of the various thermodynamic equations, it can be shown that

$$\phi_i = \exp\left[\frac{1}{RT}\int_0^\pi \left(\bar{v}_i - \frac{RT}{P}\right) dP\right]. \tag{4-41}$$

The advantage Equation 4-41 has over Equation 4-8 is that it circumvents the problem of extrapolating the pure-component equation of state deeply into the zone where the pure component exists only as a liquid. The various manipulations that have been described for the integrations involving $\bar{v}_i$ can also be performed in evaluating Equation 4-41.

Further discussion of the properties of gases and gas mixtures may be obtained from the more recent texts on thermodynamics [3–5].

4.3 LIQUID-PHASE ACTIVITY COEFFICIENTS

The determination of liquid-phase activity coefficients is not as straightforward as the determination of gas-phase activity coefficients. Although theoretically the methods developed in the preceding section are applicable to the liquid phase as well, the practical difficulties of application are formidable. The equation of state used must be capable of making the transition from gas to liquid at precisely the right pressure for mixtures as well as pure components. This requirement puts such a strain on the accuracy of the equation of state, and of the mixing rules, that a different approach than that of Section 4.2 must usually be taken.

In general the development of equations for liquid-phase activities has followed empirical or, at best, semitheoretical approaches. Consequently many equations have been proposed and undoubtedly many more will be forthcoming. Equations developed to describe the composition dependence of activity coefficients must possess certain characteristics in order to be thermodynamically possible. In particular the Gibbs–Duhem equation (Equation 2-16b) must hold at constant temperature and pressure.

Because of the thermodynamic restrictions placed on activity coefficients, it is a good practice to base a new hypothesis on the free energy of mixing and to derive the activity coefficient from the free-energy equation. By such a practice thermodynamic consistency is ensured.

The derivation of the activity coefficient from the free energy of mixing can be accomplished by defining F^M:

$$F \equiv \sum_j N_j \mu_j^\circ + F^M,$$

where F^M is known as the free energy of mixing. By differentiating at contants temperature and pressure, we get

$$\left(\frac{\partial F}{\partial N_i}\right)_{T,P,N_{j\neq i}} = \mu_i = \mu_i^\circ + \left(\frac{\partial F^M}{\partial N_i}\right)_{T,P,N_{j\neq i}}. \qquad (4\text{-}42)$$

Thus we have from Equation 2-18

$$\left(\frac{\partial F^M}{\partial N_i}\right)_{T,P,N_{j\neq i}} = RT \ln \gamma_i x_i. \qquad (4\text{-}43)$$

The free energy of mixing can be considered to be made up of the sum of the "ideal" free energy of mixing F^I and the "excess" free energy of mixing F^E. The "ideal" free energy of mixing is the value that F^M would have if the solution were ideal. Thus

$$\left(\frac{\partial F^M}{\partial N_i}\right)_{T,P,N_{j\neq i}} = \left(\frac{\partial F^I}{\partial N_i}\right)_{T,P,N_{j\neq i}} + \left(\frac{\partial F^E}{\partial N_i}\right)_{T,P,N_{j\neq i}}. \qquad (4\text{-}44)$$

By differentiating Equation 2-25, we obtain

$$\left(\frac{\partial F^I}{\partial N_i}\right)_{T,P,N_{j\neq i}} = RT \ln x_i. \qquad (4\text{-}45)$$

By substitution of Equations 4-44 and 4-45 into Equation 4-43, we get

$$RT \ln x_i + \left(\frac{\partial F^E}{\partial N_i}\right)_{T,P,N_{j\neq i}} = RT \ln \gamma_i x_i, \qquad (4\text{-}46)$$

or

$$\ln \gamma_i = \frac{1}{RT}\left(\frac{\partial F^E}{\partial N_i}\right)_{T,P,N_{j\neq i}} = \left(\frac{\partial (N_t Q)}{\partial N_i}\right)_{T,P,N_{j\neq i}}, \qquad (4\text{-}47)$$

where $Q \equiv F^E/N_t RT$.

The procedure used to deduce the activity coefficient is to first assume an equation for Q as a function of composition and then to perform the differentiation indicated in Equation 4-47. It must be noted that the differentiation is with respect to the number of moles, not the mole fraction, as independent variables. Therefore all mole fractions in the equation for Q must be converted to total moles:

$$x_i = \frac{N_i}{\sum_j N_j}. \qquad (4\text{-}48)$$

If one does not wish to make the conversion above, an alternative procedure is available. First, expand Equation 4-47:

4.3 Liquid-Phase Activity Coefficients

$$\ln \gamma_i = Q + N_t \left(\frac{\partial Q}{\partial N_i}\right)_{T,P,N_{j \neq i}}. \tag{4-49}$$

But

$$\left(\frac{\partial Q}{\partial N_i}\right)_{T,P,N_{j \neq i}} = \sum_j \left(\frac{\partial Q}{\partial x_j}\right)_{T,P,x_{k \neq j}} \frac{\partial x_j}{\partial N_i}. \tag{4-50}$$

It was previously shown that

$$\frac{\partial x_j}{\partial N_i} = -\frac{x_j}{N_t} \quad (j \neq i) \tag{4-51}$$

$$= \frac{1 - x_j}{N_t} \quad (j = i).$$

By substitution, Equation 4-49 becomes

$$\ln \gamma_i = Q + \left(\frac{\partial Q}{\partial x_i}\right)_{T,P,x_{j \neq i}} - \sum_j x_j \left(\frac{\partial Q}{\partial x_j}\right)_{T,P,x_{k \neq j}}. \tag{4-52}$$

It is implicit in the use of Equation 4-52 that all mole fractions are treated as independent variables in performing the differentiations.

The use of Equations 4-52 and 4-47 is illustrated by using the simple function

$$Q = \sum_j \sum_k A_{jk} x_j x_k,$$

where A_{jk} is a temperature-dependent parameter for the j, k pair, $A_{ii} = 0$ and $A_{jk} = A_{kj}$. The derivatives for this form are

$$\frac{\partial Q}{\partial x_j} = 2 \sum_k A_{jk} x_k.$$

Applying Equation 4-52, we have

$$\ln \gamma_i = \sum_j \sum_k A_{jk} x_j x_k + 2 \sum_j A_{ji} x_j - 2 \sum_j \sum_k A_{jk} x_j x_k,$$

$$\ln \gamma_i = 2 \sum_j A_{ji} x_j - \sum_j \sum_k A_{jk} x_j x_k.$$

For a binary system

$$\ln \gamma_1 = 2A_{21} x_2 - 2A_{21} x_1 x_2$$

$$= 2A_{21} x_2 (1 - x_1),$$

or

$$\ln \gamma_1 = 2A_{21} x_2^2 = A x_2^2. \tag{4-53}$$

This result is the simplest of the forms of equation for activity coefficients.

An alternative derivation can be obtained by direct use of Equation 4-47. The Q equation, written in terms of total moles of the components, is

$$N_t Q = \frac{1}{N_t} \sum_j \sum_k A_{jk} N_j N_k.$$

Differentiating with respect to N_i, we obtain

$$\ln \gamma_i = -\frac{1}{N_t^2} \sum_j \sum_k A_{jk} N_j N_k + \frac{2}{N_t} \sum_j A_{ji} N_j,$$

or

$$\ln \gamma_i = 2 \sum_j A_{ji} x_j - \sum_j \sum_k A_{jk} x_j x_k.$$

The reduction to a binary system is identical with the method for Equation 4-52. The choice of the method to be used is left to the reader. Both are correct, but in some instances the amount of work involved can be greatly reduced by the choice of the method.

4.3.1 Two-Constant Binary Equations

The equations derived above contain only a single constant for the calculation of the isothermal activity coefficients of a binary system. A binary system that obeys these equations is called a symmetrical system. There are very few symmetrical systems, and at least two constants are generally required for the correlation of isothermal, binary liquid-solution data. In the following paragraphs a number of the frequently used two-constant equations are described.

Hildebrand [6–10] and Scatchard [11] derived for the heat of mixing an equation that, when combined with the assumption that the entropy of mixing is ideal (regular solution), leads to a simple equation for the activity coefficient. In the Scatchard–Hildebrand equations the free-energy function is given by

$$Q = \frac{(x_1 v_1 + x_2 v_2)(\delta_1 - \delta_2)^2 \phi_1 \phi_2}{RT}, \tag{4-54}$$

where v_1, v_2 = molal volumes of pure components 1 and 2, respectively,
ϕ_1, ϕ_2 = volume fractions based on pure-component volumes,
δ = solubility parameter = $\sqrt{E_{\text{vap}}/v}$
E_{vap} = internal energy change from saturated liquid to ideal gas.

4.3 Liquid-Phase Activity Coefficients

The activity-coefficient equations can be obtained by differentiating the expression for Q:

$$\ln \gamma_1 = \frac{v_1(\delta_1 - \delta_2)^2 \phi_2^{\,2}}{RT},$$

$$\ln \gamma_2 = \frac{v_2(\delta_1 - \delta_2)^2 \phi_1^{\,2}}{RT}. \quad (4\text{-}55)$$

The two constants $v_1(\delta_1 - \delta_2)^2$ and $v_2(\delta_1 - \delta_2)^2$ can be obtained entirely from the properties of the pure components. Thus these equations are very useful in that they require no experimental solution data to determine the constants. Although they are reasonably accurate for solutions that one would expect to be regular, they generally give values of γ that are too high in the case of nonregular solutions. They are very useful in screening, however, since the activity-coefficient magnitudes for a number of solutions in a given solvent will usually be in the order predicted by the Scatchard–Hildebrand equations even though the solution is not regular. Equations 4-55 always give activity coefficients greater than unity.

The Margules [12] equations are derived by assuming

$$Q = a_{12} x_1 x_2 + a_{112} x_1^{\,2} x_2 + a_{122} x_1 x_2^{\,2}. \quad (4\text{-}56)$$

When differentiated, this free-energy function gives

$$\ln \gamma_1 = [A_{12} + 2(A_{21} - A_{12})x_1]x_2^{\,2},$$

$$\ln \gamma_2 = [A_{21} + 2(A_{12} - A_{21})x_2]x_1^{\,2}. \quad (4\text{-}57)$$

The two constants A_{12} and A_{21} are given by the infinite-dilution activity coefficients:

$$\ln \gamma_1^\infty = A_{12}$$

and

$$\ln \gamma_2^\infty = A_{21}, \quad (4\text{-}58)$$

where

$$\gamma_1^\infty \equiv \lim_{x_1 \to 0} \gamma_1.$$

The Scatchard–Hamer [13] equations have a free-energy function of the form

$$Q = a_{12} \phi_1 \phi_2 + a_{112} \phi_1^{\,2} \phi_2 + a_{122} \phi_1 \phi_2^{\,2}, \quad (4\text{-}59)$$

which gives

$$\ln \gamma_1 = \left[A_{12} + 2\left(A_{21}\frac{v_1}{v_2} - A_1\right)_2\phi_1\right]\phi_2^2,$$

$$\ln \gamma_2 = \left[A_{21} + 2\left(A_{12}\frac{v_2}{v_1} - A_{21}\right)\phi_2\right]\phi_1^2. \qquad (4\text{-}60)$$

Here, again, the constants are related to the infinite-dilution activity coefficients by $A_{12} = \ln \gamma_1^\infty$.

The Van Laar [14] equations are derivable from the form

$$Q = \frac{A_{12}A_{21}x_1x_2}{A_{12}x_1 + A_{21}x_2}. \qquad (4\text{-}61)$$

The resulting activity coefficients are given by

$$\ln \gamma_1 = A_{12}\left(\frac{A_{21}x_2}{A_{12}x_1 + A_{21}x_2}\right)^2,$$

$$\ln \gamma_2 = A_{21}\left(\frac{A_{12}x_1}{A_{12}x_1 + A_{21}x_2}\right)^2 \qquad (4\text{-}62)$$

Once again, the equation $A_{12} = \ln \gamma_1^\infty$ holds. The Van Laar equations quite frequently agree with experimental data better than either the Scatchard–Hamer or the Margules equations. However, their use is hampered somewhat by their unsuitability to systems in which A_{12} and A_{21} are of opposite sign, since the denominator becomes zero when $x_1/x_2 = |A_{21}|/|A_{12}|$. Although such systems occur infrequently and some investigators doubt the reality of their existence, there is some experimental evidence of their existence. The use of absolute values of A_{12} and A_{21} in the denominator of Equations 4-62 removes the discontinuity, but the resulting equations do not conform rigorously to the Gibbs–Duhem equations (Equations 2-16a,b) when A_{12} and A_{21} are of opposite sign. It does, however, give the same results as the original Van Laar equations when A_{12} and A_{21} are of the same sign, whether it be positive or negative.

It is possible to obtain a form that is identical with Equations 4-62 when A_{12} and A_{21} have the same sign and is thermodynamically consistent when they have opposite signs.

Equation 4-61 can be modified to

$$Q = \frac{A_{12}A_{21}x_1x_2(A_{12}x_1 + A_{21}x_2)}{(|A_{12}|x_1 + |A_{21}|x_2)^2}. \qquad (4\text{-}63)$$

Equation 4-63 is identical with Equation 4-61 when A_{12} and A_{21} have the same sign, but it maintains a positive, finite denominator when they are of opposite sign. The resulting activity-coefficient equations are

4.3 Liquid-Phase Activity Coefficients

$$\ln \gamma_1 = A_{12} Z_2^2 \left[1 + 2Z_1\left(\frac{A_{12} A_{21}}{|A_{12} A_{21}|} - 1\right)\right],$$

$$\ln \gamma_2 = A_{21} Z_1^2 \left[1 + 2Z_2\left(\frac{A_{12} A_{21}}{|A_{12} A_{21}|} - 1\right)\right],$$

(4-64)

where

$$Z_1 = \frac{|A_{12}| x_1}{|A_{12}| x_1 + |A_{21}| x_2} \quad \text{and} \quad Z_2 = 1 - Z_1.$$

Equations 4-64 are identical with the Van Laar equations when A_{12} and A_{21} are of the same sign.

All of the foregoing equations are consistent with the general form assumed by Wohl [15]:

$$Q = (2a_{12} Z_1 Z_2 + 3a_{112} Z_1^2 Z_2 + 3a_{122} Z_2^2 Z_1)(q_1 x_1 + q_2 x_2), \quad (4\text{-}65)$$

where $Z_i = q_i x_i / \sum_j q_j x_j$.

The activity-coefficient equations are

$$\ln \gamma_1 = \left[A_{12} + 2\left(A_{21}\frac{q_1}{q_2} - A_{12}\right)Z_1\right]Z_2^2,$$

$$\ln \gamma_2 = \left[A_{21} + 2\left(A_{12}\frac{q_2}{q_1} - A_{21}\right)Z_2\right]Z_1^2.$$

(4-66)

Each of the preceding equations can be obtained by making appropriate assumptions regarding the ratio q_1/q_2:

To obtain the Scatchard–Hildebrand equations:

$$\frac{q_1}{q_2} = \frac{A_{12}}{A_{21}} = \frac{v_1}{v_2}.$$

To obtain the Margules equation:

$$\frac{q_1}{q_2} = 1.$$

To obtain the Scatchard–Hamer equations:

$$\frac{q_1}{q_2} = \frac{v_1}{v_2}.$$

To obtain the Van Laar equations:

$$\frac{q_1}{q_2} = \frac{A_{12}}{A_{21}}.$$

And to obtain Equations 4-64:

$$\frac{q_1}{q_2} = \left|\frac{A_{12}}{A_{21}}\right|.$$

A recent equation of considerable merit was presented by Wilson [16]. It does not follow the Wohl concept. The free-energy function is assumed to be of the Flory–Huggins [17, 18] form, but with "local" volume fractions differing from the average volume fraction. The form is

$$Q = -[x_1 \ln(x_1 + A_{12}x_2) + x_2 \ln(x_2 + A_{21}x_1)]. \quad (4\text{-}67)$$

The resulting activity-coefficient equations are

$$\ln \gamma_1 = 1 - \ln(x_1 + A_{12}x_2) - \frac{x_1}{x_1 + A_{12}x_2} - \frac{A_{21}x_2}{x_2 + A_{21}x_1},$$

$$\ln \gamma_2 = 1 - \ln(x_2 + A_{21}x_1) - \frac{x_2}{x_2 + A_{21}x_1} - \frac{A_{12}x_1}{x_1 + A_{12}x_2}. \quad (4\text{-}68)$$

The relationship of the constants to infinite-dilution activity coefficients differs from that of equations following the Wohl form. For the Wilson equations

$$\ln \gamma_1^\infty = 1 - \ln A_{12} - A_{21},$$

$$\ln \gamma_2^\infty = 1 - \ln A_{21} - A_{12}. \quad (4\text{-}69)$$

A comparison of the above equations is discussed in greater detail in Chapter 5. In summary there is little difference between these equations for solutions that are nearly ideal or regular, although the Wilson equation seems slightly poorer in these cases. For moderate deviations from ideal or regular solution behavior the Scatchard–Hildebrand equations give activity coefficients that are quite large compared with experimental values. The other equations give comparable results when the constants are determined by curve fitting, but when evaluated by measurement of infinite-dilution activity coefficients the Van Laar and Wilson equations are definitely superior. The Wilson equation is far superior to all the others in systems that are very nonideal but do not form two liquid phases. For systems that exhibit limited miscibility no two-constant equation is adequate; the Wilson equation, which is very good as partial miscibility is approached is mathematically incapable of producing a solution to Equation 3-31.

4.3.2 Three-Constant Binary Equations

Three-constant binary equations for activity coefficients can be obtained by including additional terms in the free-energy function. Wohl [15] refers

4.3 Liquid-Phase Activity Coefficients

to his three-constant form as a "four-suffix" equation. This nomenclature arises from the fact that the constants in the free-energy function express fourth-order molecular interactions:

$$Q = (2a_{12}Z_1Z_2 + 3a_{112}Z_1{}^2Z_2 + 3a_{122}Z_2{}^2Z_1 + 4a_{1222}Z_1Z_2{}^3 \\ + 6a_{1122}Z_1{}^2Z_2{}^2 + 4a_{1112}Z_1{}^3Z_2)(q_1x_1 + q_2x_2). \quad (4\text{-}70)$$

The general equation for the activity coefficients becomes

$$\ln \gamma_1 = Z_2{}^2\left[A_{12} + 2\left(\frac{q_1}{q_2}A_{21} - A_{12} - D_{12}\right)Z_1 + 3D_{12}Z_1{}^2\right],$$

$$\ln \gamma_2 = Z_1{}^2\left[A_{21} + 2\left(\frac{q_2}{q_1}A_{12} - A_{21} - D_{21}\right)Z_2 + 3D_{21}Z_2{}^2\right], \quad (4\text{-}71)$$

where

$$D_{21} = \frac{q_2}{q_1}D_{12}.$$

If the appropriate assumptions are made, a form corresponding to each of the Wohl forms already defined is obtained:

1. The Margules form ($q_1/q_2 = 1$):

$$\ln \gamma_1 = x_2{}^2[A_{12} + 2(A_{21} - A_{12} - D)x_1 + 3Dx_1{}^2],$$

$$\ln \gamma_2 = x_1{}^2[A_{21} + 2(A_{12} - A_{21} - D)x_2 + 3Dx_2{}^2]. \quad (4\text{-}72)$$

2. The Scatchard–Hamer form ($q_1/q_2 = v_1/v_2$):

$$\ln \gamma_1 = \phi_2{}^2\left[A_{12} + 2\left(A_{21}\frac{v_1}{v_2} - A_{12} - D_{12}\right)\phi_1 + 3D_{12}\phi_1{}^2\right],$$

$$\ln \gamma_2 = \phi_1{}^2\left[A_{21} + 2\left(A_{12}\frac{v_2}{v_1} - A_{21} - D_{21}\right)\phi_2 + 3D_{21}\phi_2{}^2\right], \quad (4\text{-}73)$$

where

$$\frac{D_{12}}{D_{21}} = \frac{v_1}{v_2}.$$

3. The Van Laar form ($q_1/q_2 = A_{12}/A_{21}$):

$$\ln \gamma_1 = A_{12}Z_2{}^2\left[1 + \frac{D_{12}}{A_{12}}(3Z_1{}^2 - 2Z_1)\right],$$

$$\ln \gamma_2 = A_{21}Z_1{}^2\left[1 + \frac{D_{21}}{A_{21}}(3Z_2{}^2 - 2Z_2)\right],$$

where

$$\frac{D_{12}}{D_{21}} = \frac{A_{12}}{A_{21}}.$$

4. The alternative, absolute-value, Van Laar form ($q_1/q_2 = |A_{12}/A_{21}|$):

$$\ln \gamma_1 = A_{12} Z_2^2 \left[1 + 2Z_1 \left(\frac{A_{12} A_{21}}{|A_{12} A_{21}|} - 1 \right) + \frac{D_{12}}{A_{12}} (3Z_1^2 - 2Z_1) \right],$$

$$\ln \gamma_2 = A_{21} Z_1^2 \left[1 + 2Z_2 \left(\frac{A_{12} A_{21}}{|A_{12} A_{21}|} - 1 \right) + \frac{D_{21}}{A_{21}} (3Z_2^2 - 2Z_2) \right].$$

(4-75)

A slightly different approach was used by Redlich and Kister [19], who assumed the free-energy function to be of the form

$$Q = x_1 x_2 [B' + C'(x_1 - x_2) + D'(x_1 - x_2)^2 + \cdots]. \quad (4\text{-}76)$$

In their three-constant form the Redlich–Kister equations are

$$\ln \gamma_1 = x_2^2 [B' + C'(3x_1 - x_2) + D'(x_1 - x_2)(5x_1 - x_2)],$$

$$\ln \gamma_2 = x_1^2 [B' + C'(x_1 - 3x_2) + D'(x_1 - x_2)(x_1 - 5x_2)].$$

(4-77)

In the two-constant form the Redlich–Kister equations are identical with the Margules equations, but for higher order terms the method of truncation is different from that used in the Wohl method.

Black [20] assumed the function Q to consist of two contributions:

$$Q = Q_v + Q_e, \quad (4\text{-}78)$$

where Q_v is assumed to be the Van Laar two-constant form for Q and Q_e is given by

$$Q_e = C_{12}(x_1 - x_2)^2 x_1 x_2. \quad (4\text{-}79)$$

The activity coefficients derived are

$$\ln \gamma_1 = A_{12} \left(\frac{A_{21} x_2}{A_{12} x_1 + A_{21} x_2} \right)^2 + C_{12} x_2^2 (2x_2 - 1)(6x_2 - 5),$$

$$\ln \gamma_2 = A_{21} \left(\frac{A_{12} x_1}{A_{12} x_1 + A_{21} x_2} \right)^2 + C_{12} x_1^2 (2x_1 - 1)(6x_1 - 5).$$

(4-80)

Another approach was taken by R. H. Fariss,† who assumed a modified Van Laar form:

† Private communication to the author, 1963.

4.3 Liquid-Phase Activity Coefficients

$$Q = \frac{A_{12}A_{21}x_1x_2}{(A_{12}x_1 + A_{21}x_2)(1 + \beta x_1 x_2)}. \tag{4-81}$$

The activity coefficients are given by

$$\ln \gamma_1 = \frac{A_{12}Z_2^2}{1 + \beta x_1 x_2}\left\{1 + \frac{\beta x_1(2x_1 - 1)[(A_{12}/A_{21})x_1 + x_2]}{1 + \beta x_1 x_2}\right\},$$

$$\ln \gamma_2 = \frac{A_{21}Z_2^2}{1 + \beta x_1 x_2}\left\{1 + \frac{\beta x_2(2x_2 - 1)[(A_{21}/A_{12})x_1 + x_2]}{1 + \beta x_1 x_2}\right\}, \tag{4-82}$$

where

$$Z_1 = \frac{A_{12}x_1}{A_{12}x_1 + A_{21}x_2}; \quad Z_2 = 1 - Z_1.$$

These equations are identical with the Van Laar two-constant equations if $\beta = 0$. The discontinuity for A_{12} and A_{21} having opposite signs occurs here also; it can be removed in the same manner as was done in Equation 4-63 to give for the activity coefficients

$$\ln \gamma_1 = \frac{A_{12}Z_2^2}{1 + \beta x_1 x_2}\left\{1 + 2Z_1\left(\frac{|A_{12}A_{21}|}{A_{12}A_{21}} - 1\right)\right.$$
$$\left. + \frac{\beta x_1(2x_1 - 1)[(A_{12}/A_{21})x_1 + x_2]}{1 + \beta x_1 x_2}\right\},$$

$$\ln \gamma_2 = \frac{A_{21}Z_1^2}{1 + \beta x_1 x_2}\left\{1 + 2Z_2\left(\frac{|A_{12}A_{21}|}{A_{12}A_{21}} - 1\right)\right. \tag{4-83}$$
$$\left. + \frac{\beta x_2(2x_2 - 1)[(A_{21}/A_{12})x_2 + x_1]}{1 + \beta x_1 x_2}\right\}.$$

Other equations have been proposed, and others will undoubtedly be proposed in the future. A new equation presented while this book was being written, known as the NRTL equation [21], promises to be quite significant.

Of the three-constant equations given, the Wohl forms and the Fariss form have the advantage that the two constants A_{12} and A_{21} are determined uniquely by the infinite-dilution activity coefficients:

$$\ln \gamma_1^\infty = A_{12}; \quad \ln \gamma_2^\infty = A_{21}.$$

For these forms the third-constant terms vanish when $x_1 = 1$ or $x_1 = 0$. This is a distinct advantage since it is possible in some cases to measure infinite-dilution activity coefficients directly, and some of the most difficult separation problems occur in the dilute region.

4.3.3 Multicomponent Equations

The multicomponent equations are, generally, extensions of the binary forms.

The Scatchard–Hildebrand equations can be adapted to multicomponents by a method suggested by Hildebrand and further exploited by Chao and Seader [22]. The multicomponent system is considered to be a pseudobinary consisting of the component whose activity coefficient is to be calculated at infinite dilution in the total system. The activity coefficient of component i then becomes

$$\ln \gamma_i = \frac{v_i(\delta_i - \bar{\delta})^2}{RT}, \tag{4-84}$$

where

$$\bar{\delta} = \sum_j \phi_j \delta_j.$$

It is a consequence of regular-solution theory that the quantity $v_i(\delta_i - \bar{\delta})^2$ should be independent of temperature; thus, a value obtained at any given temperature may be used at any other temperature.

The equations derivable from the Wohl formulation giving two constants for the binary case can be represented by a multicomponent form of the free-energy function as follows:

$$Q = \left(\sum_i q_i x_i\right)\left(\sum_i \sum_j a_{ij} Z_i Z_j + \sum_i \sum_j \sum_k a_{ijk} Z_i Z_j Z_k\right). \tag{4-85}$$

The first summation can be altered by the multiplication

$$\sum_i \sum_j a_{ij} Z_i Z_j = \sum_i \sum_j a_{ij} Z_i Z_j \sum_k Z_k,$$

since $\sum_k Z_k = 1$. The term then becomes

$$\sum_i \sum_j a_{ij} Z_i Z_j = \sum_i \sum_j \sum_k a_{ij} Z_i Z_j Z_k.$$

If equivalent forms are used, it is also true that

$$\sum_i \sum_j \sum_k a_{ij} Z_i Z_j Z_k = \sum_i \sum_j \sum_k a_{ik} Z_i Z_j Z_k = \sum_i \sum_j \sum_k a_{jk} Z_i Z_j Z_k.$$

By averaging the three forms, we can obtain

$$\sum_i \sum_j a_{ij} Z_i Z_j = \sum_i \sum_j \sum_k [\tfrac{1}{3}(a_{ij} + a_{ik} + a_{jk})] Z_i Z_j Z_k$$

4.3 Liquid-Phase Activity Coefficients

This leads to a more compact form of the Q-function:

$$Q = \left(\sum_i q_i x_i\right) \sum_i \sum_j \sum_k [\tfrac{1}{3}(a_{ij} + a_{ik} + a_{jk}) + a_{ijk}] Z_i Z_j Z_k$$

$$= \left(\sum_i q_i x_i\right) \sum_i \sum_j \sum_k \beta_{ijk} Z_i Z_j Z_k. \tag{4-86}$$

It is assumed in the Wohl formulation that coefficients with identical subscripts except for their order are equal and that the coefficients with *all* subscripts equal are zero.
Thus

$$a_{ij} = a_{ji},$$

$$a_{ii} = 0,$$

$$a_{ijk} = a_{ikj} = a_{jki} = a_{jik} = a_{kij} = a_{kji},$$

$$a_{ijj} \neq 0 \quad (\text{for } i \neq j),$$

$$a_{iii} = 0.$$

These conditions lead to a similar situation for the β-coefficients:

$$\beta_{ijk} = \beta_{ikj} = \beta_{kij} = \beta_{kji} = \beta_{jik} = \beta_{jki},$$

$$\beta_{iij} \neq 0 \quad (\text{for } i \neq j),$$

$$\beta_{iii} = 0.$$

When Equation 4-86 is differentiated to give the activity coefficient of component m, we obtain

$$\ln \gamma_m = q_m \left(3 \sum_j \sum_k \beta_{mjk} Z_j Z_k - 2 \sum_i \sum_j \sum_k \beta_{ijk} Z_i Z_j Z_k\right). \tag{4-87}$$

It is desirable to obtain equations that involve only the ratios of the q terms. This can be done by defining a new coefficient:

$$C_{ijk} = (q_i + q_j + q_k)\beta_{ijk}. \tag{4-88}$$

The C terms will maintain the properties of commutation of indices inherent in the β terms and give for the activity coefficient

$$\ln \gamma_m = 3 \sum_j \sum_k \frac{C_{mjk} Z_j Z_k}{\left(1 + \dfrac{q_j}{q_m} + \dfrac{q_k}{q_m}\right)} - 2 \sum_i \sum_j \sum_k \frac{C_{ijk} Z_i Z_j Z_k}{\left(\dfrac{q_i}{q_m} + \dfrac{q_j}{q_m} + \dfrac{q_k}{q_m}\right)}. \tag{4-89}$$

Coefficients in which two of the subscripts are equal are related to the binary coefficients by the equation

$$C_{ijj} = C_{jij} = C_{jji} = \frac{A_{ij}}{3}\left(1 + 2\frac{q_j}{q_i}\right). \tag{4-90}$$

Coefficients whose subscripts are all identical are zero: $C_{iii} = 0$. There are, unfortunately, terms in which the three subscripts are all different. In a ternary system there is only one such coefficient, but in the quarternary there are three, and in a five component system there are ten. These coefficients cannot be obtained from data on the binary systems alone. Consequently the ternary constants, C_{ijk}, with i, j, and k different, are rarely known and are usually assumed to have some known value. Two possibilities are often used: (a) the C_{ijk} may be assumed zero or (b) since they may be shown equal to

$$\frac{1}{12}\left[\frac{A_{ij}}{q_i} + \frac{A_{ji}}{q_j} + \frac{A_{ik}}{q_i} + \frac{A_{ki}}{q_k} + \frac{A_{jk}}{q_j} + \frac{A_{kj}}{q_k}\right.$$

$$\left. - 3(a_{iij} + a_{iik} + a_{jjk} + a_{kki} + a_{jkk} + a_{jii})\right] + a_{ijk},$$

the term C_{ijk} is given the value of the six terms containing the A terms. It is apparent that there are many options available in ordering the coefficients.

The assumptions regarding q-ratios for the Wohl equations can be applied to give multicomponent equations for the Margules, Scatchard–Hamer, and Van Laar equations, as was done for the binary systems. The Van Laar equation poses a particularly sticky problem, however, since the assumptions are made that

$$\frac{q_i}{q_j} = \frac{A_{ij}}{A_{ji}}; \quad \frac{q_i}{q_k} = \frac{A_{ik}}{A_{ki}}; \quad \frac{q_j}{q_k} = \frac{A_{jk}}{A_{kj}},$$

and so on. However, q_i/q_j can be computed by the equation

$$\frac{q_i}{q_j} = \frac{q_i\, q_k}{q_k\, q_j} = \frac{A_{ik}\, A_{kj}}{A_{ki}\, A_{jk}}.$$

This gives the restrictive requirement

$$\frac{A_{ij}}{A_{ji}} = \frac{A_{ik}\, A_{kj}}{A_{ki}\, A_{jk}}. \tag{4-91}$$

Since Equation 4-91 is not generally true of the constituent binary systems, the Margules equation has usually been used instead of the Van Laar for multicomponent calculations. This is an unfortunate effect since the Van Laar form so often gives a better representation of the binaries than does the Margules equation.

4.3 Liquid-Phase Activity Coefficients

The Wilson equation does not introduce any specific ternary constants in its multicomponent form. The free-energy function is given by

$$Q = -\sum_i x_i \ln\left(\sum_j A_{ij} x_j\right), \qquad (4\text{-}92)$$

where the A_{ij} terms are the A_{ij} values for the i,j binary system and $A_{ii} = 1$. The differentiation gives for the activity coefficient

$$\ln \gamma_i = 1 - \ln\left(\sum_j A_{ij} x_j\right) - \sum_j \frac{A_{ji} x_j}{\sum_k A_{jk} x_k}. \qquad (4\text{-}93)$$

The equations giving three constants in the binary form can also be extended to the general multicomponent case. The free-energy function for the Wohl formulation in this case is

$$\frac{Q}{\sum_i x_i q_i} = \sum_i \sum_j a_{ij} Z_i Z_j + \sum_i \sum_j \sum_k a_{ijk} Z_i Z_j Z_k + \sum_i \sum_j \sum_k \sum_l a_{ijkl} Z_i Z_j Z_k Z_l.$$

This can be transformed to the more concise form

$$Q = \left(\sum_i \sum_j \sum_k \sum_l \beta'_{ijkl} Z_i Z_j Z_k Z_l\right) \sum_i q_i x_i,$$

where

$$\beta'_{ijkl} = a_{ijkl} + \tfrac{1}{4}(\beta_{ijk} + \beta_{ikl} + \beta_{ijl} + \beta_{jkl})$$

and β_{ijk} has the same meaning as in Equation 4-85. In order that activity coefficients be defined in terms of ratios of the q terms rather than the q terms themselves, the substitution $E_{ijkl} = (q_i + q_j + q_k + q_l)\beta'_{ijkl}$ is made, giving

$$Q = \left(\sum_i q_i x_i\right) \sum_i \sum_j \sum_k \sum_l \frac{E_{ijkl} Z_i Z_j Z_k Z_l}{q_i + q_j + q_k + q_l}. \qquad (4\text{-}94)$$

After the appropriate differentiation, the activity coefficient is obtained:

$$\ln \gamma_m = 4 \sum_i \sum_j \sum_k \frac{E_{ijkm} Z_i Z_j Z_k}{[1 + (q_i/q_m) + (q_j/q_m) + (q_k/q_m)]}$$

$$- 3 \sum_i \sum_j \sum_k \sum_l \frac{E_{ijkl} Z_i Z_j Z_k Z_l}{[(q_i/q_m) + (q_j/q_m) + (q_k/q_m) + (q_l/q_m)]}. \qquad (4\text{-}95)$$

Again, forms can be obtained from the Wohl equation corresponding to the Margules ($q_i/q_j = 1$), Scatchard–Hamer ($q_i/q_j = v_i/v_j$), or Van Laar forms ($q_i/q_j = A_{ij}/A_{ji}$); and the restrictions $A_{ij}/A_{ji} = (A_{ik}/A_{ki})(A_{kj}/A_{jk})$ are once again imposed on the Van Laar form. The E terms involving only two components can be determined from binary system information:

$$E_{ijjj} = \frac{A_{ij}}{4}\left(1 + 3\frac{q_j}{q_i}\right), \tag{4-96}$$

$$E_{iijj} = \tfrac{1}{3}\left(1 + \frac{q_j}{q_i}\right)\left(\frac{q_i}{q_j}A_{ji} + A_{ij} - D_{ij}\right). \tag{4-97}$$

The order of the subscripts is immaterial:

$$E_{ijkl} = E_{ikjl} = E_{lkij},$$

and so on, and four identical subscripts designate a zero value for the coefficient:

$$E_{iiii} = 0.$$

In this form of the Wohl equation there are three specific ternary constants corresponding to each three-component combination (E_{iijk}, E_{ijjk}, and E_{ijkk}), and one specific quaternary constant (E_{ijkl}) for every four-component combination. Sufficient data for this many multicomponent constants are rarely available; consequently these equations are rarely used without assigning zero values to the ternary and quaternary constants.

The Fariss equation, in its multicomponent form, has two interesting features that are missing from the Wohl formulation; it introduces no specific ternary or higher order constants and preserves the binary Van Laar form without ambiguity. The multicomponent free-energy function is given by

$$Q = \tfrac{1}{2}\sum_i\sum_j \frac{\alpha_{ij}x_i x_j}{1 + \beta_{ij}x_i x_j}, \tag{4-98}$$

where

$$\alpha_{ij} = \frac{2A_{ij}A_{ji}}{A_{ji}(1 + x_j - x_i) + A_{ij}(1 - x_j + x_i)},$$

$$A_{ij} = \ln \gamma_i^\infty \quad \text{(in pure } j\text{)},$$

$$A_{ii} = 0,$$

$$\beta_{ij} = \beta_{ji}.$$

The activity coefficient derived from this form is

$$\ln \gamma_m = Q + \tfrac{1}{2}\Biggl\{\sum_i \left[\frac{x_i(\alpha_{im} - Q_{im}\beta_{im})}{1 + \beta_{im}x_i x_m} + \frac{Q_{im}\alpha_{im}(A_{im} - A_{mi})}{2A_{im}A_{mi}}\right] \\ - \sum_i\sum_j x_j\left[\frac{x_i(\alpha_{ij} - Q_{ij}\beta_{ij})}{1 + \beta_{ij}x_i x_j} + \frac{Q_{ij}\alpha_{ij}(A_{ij} - A_{ji})}{2A_{ij}A_{ji}}\right]\Biggr\}, \tag{4-99}$$

4.3 Liquid-Phase Activity Coefficients

where

$$Q_{ij} = \frac{\alpha_{ij} x_i x_j}{1 + \beta_{ij} x_i x_j},$$

and

$$\alpha_{ij} = \frac{2 A_{ij} A_{ji}}{(A_{ij} + A_{ji}) + (x_j - x_i)(A_{ji} - A_{ij})}.$$

When $\beta_{ij} = 0$, the Van Laar form is obtained in the binary, with the resultant multicomponent form

$$\ln \gamma_m = Q + \tfrac{1}{2} \left\{ \sum_i \alpha_{im} \left[x_i + \frac{Q_{im}(A_{im} - A_{mi})}{2 A_{im} A_{mi}} \right] \right.$$
$$\left. - \sum_i \sum_j \alpha_{ij} x_j \left[x_i + \frac{Q_{ij}(A_{ij} - A_{ji})}{2 A_{ij} A_{ji}} \right] \right\}. \quad (4\text{-}100)$$

The Black Q_e function in multicomponent form is

$$Q_e = \tfrac{1}{2} \sum_i \sum_j C_{ij}(x_i - x_j)^2 x_i x_j, \quad (4\text{-}101)$$

which gives, when differentiated,

$$\ln (\gamma_i)^e = \sum_j C_{ij} x_j (x_i - x_j)^2 + 2 \sum_j C_{ij} x_i x_j (x_i - x_j)$$
$$- \tfrac{3}{2} \sum_j \sum_k C_{jk} x_j x_k (x_j - x_k)^2, \quad (4\text{-}102)$$

where $C_{ij} = C_{ji}$ and $C_{ii} = 0$.

The value of γ_i is obtained from

$$\ln \gamma_i = \ln(\gamma_i)^v + \ln (\gamma_i)^e, \quad (4\text{-}103)$$

where $\ln(\gamma_i)^v$ is identical with the two-suffix Van Laar equation value for $\ln \gamma_i$.

As yet, there has been no comprehensive comparison test of these various equations on multicomponent data; however, we would expect that the comments regarding binary systems would carry over to multicomponent systems. The Wohl method, however, has the potential of better data fitting simply by the presence of additional empirical constants specific to the multicomponent combinations. Since sufficient data for the irdetermination are rarely available, this potential is largely academic. The Wohl, Wilson, and Fariss equations have constants that can be determined entirely from activity coefficients at infinite dilution; since there are some correlations and means of measurement for infinite-dilution activity coefficients, this can be an important advantage. The Black modification does not have constants determined entirely from dilute-solution activity coefficients.

For a number of years Monsanto process engineers have been successfully using a modification of the Van Laar equations in process simulations. The method overcomes the ambiguity of the Van Laar equations in multicomponent systems and the discontinuity in systems with mixed deviations. The computations for multicomponent systems are considerably less complex than the computations required for the other types of equations discussed. The multicomponent computation uses an averaging method that takes some liberties with regard to thermodynamic consistency, but this has caused no serious difficulty in the simulations.

The four-suffix equation for a binary system is given by

$$\ln \gamma_1 = A_{12} Z_2^2 \left[1 + 2Z_1 \left(\frac{|A_{12} A_{21}|}{A_{12} A_{21}} - 1 \right) + D'_{12}(Z_1^2 - \tfrac{2}{3} Z_1) \right], \quad (4\text{-}104)$$

where

$$Z_1 = \frac{|A_{12}| x_1}{|A_{12}| x_1 + |A_{21}| x_2}; \quad D'_{12} = D'_{21}.$$

The equation for γ_2 can be obtained by interchanging subscripts. If we assume each component to exist in a pseudobinary system with the form of Equation 4-104, we have

$$\ln \gamma_i = A_i (1 - Z_i)^2 \left[1 + 2Z_i \left(\frac{|A_i B_i|}{A_i B_i} - 1 \right) + D'_i(Z_i^2 - \tfrac{2}{3} Z_i) \right], \quad (4\text{-}105)$$

where

$$Z_i = \frac{|A_i| x_i}{|A_i| x_i + |B_i|(1 - x_i)}.$$

The values of A_i, B_i, and D'_i are determined by averaging as follows:

$$A_i = \frac{\sum_{j \neq i} A_{ij} x_j^N}{(1 - x_i)^N},$$

$$B_i = \frac{\sum_{j \neq i} A_{ji} x_j^N}{(1 - x_i)^N},$$

$$D'_i = \frac{\sum_{j \neq i} D'_{ij} x_j^N}{(1 - x_i)^N}.$$

A value of unity is usually assigned to N, although N can be adjusted for best fit to experimental multicomponent data if these are available. The term involving $A_i B_i$ or $A_{12} A_{21}$ is necessary only when the two factors have

4.3 Liquid-Phase Activity Coefficients

opposite signs. The term involving D' is normally used only when fairly large positive deviations are encountered but the mutual solubility of the components is appreciable.

Several characteristics of the pseudobinary equations are noteworthy:

1. If the system is a binary and A_{ij} and A_{ji} have the same sign, the equations are identical with four-suffix Van Laar equations.
2. If $D' = 0$, the equations become the three-suffix Van Laar equations.
3. If N is assigned, only binary-system parameters are necessary to define the multicomponent system.
4. If N is adjustable, an improved fit of multicomponent data can be obtained.
5. The equations are computationally simple.
6. The term A_{ij} is uniquely defined by the value of γ_i^∞ in component j as solvent: $A_{ij} = \ln \gamma_i^\infty$ in component j.
7. Completely miscible and partially miscible systems can be represented.
8. Mixed deviations present no problem.

Probably the most significant characteristic of these equations, however, is the fact that they have been used successfully within Monsanto to simulate the operation of hundreds of sets of operating conditions of actual columns, including ordinary distillations, extractive distillations, and heterogeneous and homogeneous azeotropic distillations. The results have been applied directly to design, rating, and troubleshooting problems.

4.3.4 Correlations for Infinite-Dilution Activity Coefficients

For any of the equations giving only two constants in the binary form a knowledge of the infinite-dilution activity coefficient γ^∞ for each component of the binary uniquely determines the entire composition dependence of the activity coefficients of the binary. Since these same constants become a part of the multicomponent equations and in some cases are the only constants in the multicomponent equations, the ability to predict γ^∞ is an invaluable aid to phase-equilibrium calculations. In equations giving three constants the equation relating γ^∞ to the constants reduces the number to be determined by curve fitting to one.

Aside from the statistical significance of the value of γ^∞, it is a characteristic of almost all nonideal solutions that the greatest deviation from ideal-solution behavior occurs at infinite dilution. It is a characteristic of separation or purification processes that the most costly portion of the purification occurs in the region approaching infinite dilution where relatively pure material is converted to material meeting specifications of very high purity.

For the reasons cited above, correlations aimed at the prediction of phase equilibrium are, and should be, concentrated toward correlation of γ^∞, the infinite-dilution activity coefficient. It is a corollary to this statement that experimental measurements should also be weighted heavily in the region of dilute solutions.

The simplest of the correlations available is the equation of Hildebrand [9] for the activity coefficient of component 1 at infinite dilution in solvent, component 2:

$$\ln \gamma_1^\infty = \frac{v_1(\delta_1 - \delta_2)^2}{RT}. \tag{4-106}$$

This equation is applicable to regular solutions and has found wide application in systems consisting of hydrocarbons. It is not good for systems involving polar or hydrogen-bonding components.

For systems in which there is no heat of mixing but for which there is an effect of volume on the entropy of mixing the Flory–Huggins [17, 18] form can be applied:

$$\ln \gamma_1^\infty = \ln \frac{v_1}{v_2} + 1 - \frac{v_1}{v_2}. \tag{4-107}$$

In systems in which the *only* deviation from regular-solution behavior is due to volume, Equations 4-106 and 4-107 may be combined:

$$\ln \gamma_1^\infty = 1 + \ln \frac{v_1}{v_2} - \frac{v_1}{v_2} + \frac{v_1(\delta_1 - \delta_2)^2}{RT}. \tag{4-108}$$

Equations 4-106 and 4-107 both tend to predict too great a deviation where they do not apply. Since their effects are often of opposite sign, one cannot predict the direction of the error in Equation 4-108.

Because of the difficulty in handling polar systems, Hildebrand [9] has suggested and Blanks and Prausnitz [23] have further developed the idea of splitting the solubility parameter into two parts:

$$\delta^2 = \lambda^2 + \tau^2, \tag{4-109}$$

where λ is the solubility parameter ascribed to nonpolar effects and τ is the polar solubility parameter. The value of λ for a polar component is set equal to the solubility parameter of a nonpolar molecule of the same size and shape, and at the same reduced temperature, as the polar molecule. The value of τ is obtained by solving Equation 4-109. The nonpolar molecule used in this context is called the homomorph of the polar molecule. Blanks and Prausnitz consider two cases. If component 1 is polar and component 2 nonpolar, γ^∞ is, in the absence of volume effects on the entropy of mixing, given by

4.3 Liquid-Phase Activity Coefficients

$$\ln \gamma_1^\infty = \frac{v_1}{RT}\left[(\lambda_1 - \delta_2)^2 + \frac{T_*}{T}(\tau_1^2 - 2\psi)\right],$$

$$\ln \gamma_2^\infty = \frac{v_2}{RT}\left[(\lambda_1 - \delta_2)^2 + \frac{T_*}{T}(\tau_1^2 - 2\psi)\right],$$

(4-110)

where ψ is an empirical parameter that Blanks and Prausnitz describe as a function of $\lambda_2\tau_1$, and T_* is the temperature at which τ_1 was calculated. If both components are polar, γ^∞ is given by

$$\ln \gamma_1^\infty = \frac{v_1}{RT}\left[(\lambda_1 - \lambda_2)^2 + \frac{T_*}{T}(\tau_1 - \tau_2)^2\right]. \tag{4-111}$$

It should be noted that in Equation 4-111 subscripts can be interchanged to calculate γ_2^∞, whereas they cannot in Equation 4-110. Weimer and Prausnitz [24] have developed a comprehensive correlation for a number of hydrocarbon families in many polar solvents based on the homomorph concept. A further extension has been made by Helpinstill and Van Winkle [25].

Pierotti, Deal, and Derr [26] have published a rather extensive correlation of values of γ^∞ for homologous series of hydrocarbons in various specific solvents or in some cases in a homologous series of solvents. In this application a homologous series is considered to be a series of compounds that differ from each other only by the number of methylene groups ($-CH_2-$) connected in a straight-chain configuration. For a homologous series of solvents their correlation takes the form

$$\log \gamma_1^\infty = A_{12} + \frac{B_2 n_1}{n_2} + \frac{C_1}{n_1} + D(n_1 - n_2)^2 + \frac{F_2}{n_2}, \tag{4-112}$$

where n_1 and n_2 = numbers of carbon atoms in the methylene chain,

A_{12} = a coefficient that depends on the nature of the functional groups of each of the components (the functional group is the part of the molecule that is common to all the members of a series),

B_2 = a coefficient determined by the functional group of the solvent series (component 2),

C_1 = a coefficient determined by the functional group of the solute series (component 1),

D = a coefficient independent of the functional groups,

F_2 = a coefficient determined by the solvent's functional group.

If the solvent is fixed, rather than a member of a series, the equation becomes

$$\log \gamma_1^\infty = K_{12} + B_2' n_1 + \frac{C_1}{n_1} + D(n_1 - n_2)^2, \tag{4-113}$$

where

$$K_{12} = A_{12} + \frac{F_2}{n_2}; \quad B'_2 = \frac{B_2}{n_2}.$$

All of the coefficients are temperature dependent, and Pierotti et al. [26] have published an extensive table of these coefficients at various specific temperatures for the homologous series that they have correlated. By means of special counting rules described in their paper it is also possible to include a certain degree of branching in the chain and the addition of repeating cyclic units as though they were parts of a homologous series. The authors claim an overall average deviation of 8% in the calculation of γ^∞.

Null and Palmer [27] have combined the basic ideas of Weimer and Prausnitz [24] and of Helpinstill and Van Winkle [25] with those of Wiehe and Bagley [28]. The basic concept of their correlation is that the various contributions to $\ln \gamma_1^\infty$, are additive:

$$\ln \gamma_1^\infty = (\ln \gamma_1^\infty)_\text{I} + (\ln \gamma_1^\infty)_\text{II} + (\ln \gamma_1^\infty)_\text{III} + \cdots,$$

where the roman numeral subscripts refer to various effects or contributions. The effects included are the following:

1. Scatchard–Hildebrand regular-solution heat of mixing.
2. Polar-energy and induced-dipole effect as described by Weimer and Prausnitz [24] and by Helpinstill and Van Winkle [25].
3. Entropy effects due to the Wiehe and Bagley [28] concept of an effective polymerization equilibrium. This reduces to the Flory–Huggins [17, 18] terms for nonassociating molecules.
4. Free energy of the broken polymer bonds as the associating component goes into solution.

Other effects, such as cross-association between two components, can be included later as they are developed. The effects included by Weimer and Prausnitz, and by Helpinstill and Van Winkle, are expressed as

$$(\ln \gamma_1^\infty)_\text{polar} = \frac{v_1}{RT} [(\lambda_1 - \lambda_2)^2 + \eta_{12}(\tau_1 - \tau_2)^2] + \ln \frac{v_1}{v_2} + 1 - \frac{v_1}{v_2}, \tag{4-114}$$

where η_{12} is correlated by

$$\eta_{12} = d - e(\lambda_1 - \lambda_2)^2, \tag{4-115}$$

d and e being the characteristic parameters of the *solute*. The complete correlation is given in Table 4.1. One merely obtains $\ln \gamma^\infty$ by adding the appropriate terms in the table. Instructions for including or excluding each

Table 4.1 Additive Contribution Terms to $\ln \gamma_1^\infty$

Description[a]	Contribution to $\ln \gamma_1^\infty$	Include if
Athermal mixing, component 1	$1 + \ln \dfrac{v_1}{v_2} + \dfrac{\ln K_1}{K_1}[\ln(1+K_1) - K_1]$	Component 1 is an associating compound
	$1 + \ln \dfrac{v_1}{v_2}$	Component 1 is nonassociating
Athermal mixing, component 2	$-\dfrac{v_1}{v_2}\left\{\dfrac{1}{1+K_2} + \dfrac{\ln K_2}{K_2} \cdot \left[\ln(1+K_2) - \dfrac{K_2}{1+K_2}\right]\right\}$	Component 2 is associating
	$-\dfrac{v_1}{v_2}$	Component 2 is nonassociating
Free energy of association bonds broken	$\dfrac{-\Delta g_1}{RT}\left[1 - \dfrac{\ln(1+K_1)}{K_1}\right]$	Component 1 is associating
	$\dfrac{-\Delta g_2}{RT}\left[\dfrac{\ln(1+K_2)}{K_2} - \dfrac{1}{1+K_2}\right]\dfrac{v_1}{v_2}$	Component 2 is associating
Heat of mixing	$\dfrac{v_1}{RT}[(\lambda_1 - \lambda_2)^2 + \eta_{12}(\tau_1 - \tau_2)^2]$	Both components polar
	$\dfrac{v_1}{RT}[(\lambda_1 - \lambda_2)^2 + \eta_{12}\tau_2^2]$	Component 1 nonpolar, component 2 polar
	$\dfrac{v_1}{RT}[(\lambda_1 - \lambda_2)^2 + \eta_{12}\tau_1^2]$	Component 1 polar, component 2 nonpolar
	$\dfrac{v_1}{RT}(\lambda_1 - \lambda_2)^2$	Neither component polar

[a] Component 1 is the solute Component 2 is the solvent.

term are given under the column labeled "Include if." The value of λ is determined by regression analysis of γ^∞ data for the new correlation, rather than from homomorph plots. In the new correlation the usual solubility parameter is expressed as

$$\delta^2 = \frac{\Delta U_{vap}}{v} = \lambda^2 + \tau^2 + \zeta^2, \qquad (4\text{-}116)$$

where λ^2 represents the regular-solution energy density, τ^2 represents the polar energy density, and ζ^2 represents the energy density due to association.

For use in the correlation δ is determined as

$$\delta^2 = \frac{2.303RB}{v}\left(\frac{T}{t+C}\right)^2 - \frac{RT}{v}, \qquad (4\text{-}117)$$

where B and C are constants of the Antoine vapor-pressure equation:

$$\log p^\circ = A - \frac{B}{t+C}. \qquad (4\text{-}118)$$

The values of λ are correlated from experimental γ^∞ data by a similar type of equation:

$$\lambda^2 = \frac{2.303R\beta_\lambda}{v}\left(\frac{T}{t+T_L}\right)^2 - \frac{RT}{v}. \qquad (4\text{-}119)$$

The association-energy density is given by

$$\zeta^2 = -\frac{\Delta h}{v}\left[\frac{K - \ln(1+K)}{K}\right], \qquad (4\text{-}120)$$

where

$$K = \exp\left(1 - \frac{\Delta s}{R} + \frac{\Delta h}{RT}\right). \qquad (4\text{-}121)$$

The terms Δs and Δh are the entropy and heat of association, respectively, as defined by Wiehe and Bagley [28]. The polar energy density is obtained by difference:

$$\tau^2 = \delta^2 - \lambda^2 - \zeta^2. \qquad (4\text{-}122)$$

Thus, in addition to the usual thermodynamic properties, each component is characterized by the parameters β_λ, T_L, Δs, Δh, d, and e. The last two

(d and e) are used only for the solute. The molar volume v is calculated by

$$v = \frac{M}{a - bt}.$$

The quantity Δg in Table 4.1 is given by

$$\Delta g = \Delta h - T\Delta s.$$

Parameters for 84 compounds were determined by regression analysis of the γ^∞ data of the Shell Development Company [26, 29, 30] and of Gerster [31]. These parameters are listed in Table 4.2. The Antoine constants are for $\log_{10} p°$ (millimeters of mercury) and temperature in degrees Celsius. The abbreviations of component names are listed in Table 4.3. The overall average error in prediction for 845 literature data points (approximately 300 systems) was 25% in γ^∞, equivalent to a standard deviation of 41%. This value is somewhat misleading, however, since elimination of the data for chrysene, terphenyl, and tricontane (46 points) reduces the average error to 21.6%, whereas the average error for all saturated hydrocarbons (including cyclic ones) in all solvents was less than 9% (standard deviation 14%). The errors are, generally, greater when the polar or associating component is the solute than when it is the solvent. The correlation gives values that are quite adequate for screening purposes and that in some cases can be used for design.

In general the correlations are quite helpful for screening purposes or for calculating the distribution of minor components in a separation process. For major components in a highly nonideal solution or for minor components that are critical to a product specification experimental data on which to base the values of the parameters for composition dependence of activity coefficients are essential to reliable process design.

4.4 SOLID-PHASE ACTIVITY COEFFICIENTS

Far less work has been directed to activity-coefficient equations for the solid phase than for either the gas or liquid phase. This is undoubtedly due to the difficulty in obtaining experimental data that give the value of activity coefficients accurately in solid-state solutions.

Generally the approach has been to use one of the liquid-phase equations. The symmetrical Equation 4-52 has probably been used more than any other, although there is no particular reason to believe that the solid phase obeys this symmetrical equation any better than the liquid phase.

It is the author's opinion that, if activity-coefficient equations are investigated extensively for the solid phase, essentially the same equations as are used for the liquid phase will apply to similar circumstances in the solid phase.

Table 4.2 Parameters for Use in the γ^∞ Correlation

COMP.	TYPE	ANTOINE CONSTANTS A_{VP}	B_{VP}	C	MOL. WT. M	DENS. COEFFS. a	b	c	B_1	T_L	Δs	Δh	d	e
ACETONE	12	7.23967	1267.260	234.690	58.080	0.81283	0.001113	1591.699	1117.730	13.4981	3287.800	0.325090	-0.046216
AC-NITRL	12	7.11988	1314.400	230.000	41.050	0.80730	0.001080	877.482	210.337	-4.4430	587.676		
AC-PHENO			1946.100	220.000	120.180	1.03536	0.000863	2121.486	237.137				
ANILINE		7.24170	1675.300	200.000	93.120	1.03865	0.000846	1309.310	182.865				
ANTHRAC	3		1287.800	143.486	178.220	1.27349	0.000870	1287.800	143.486				
BENZENE	3	6.90565	1211.030	220.790	78.110	0.89964	0.001074	1211.030	220.790			0.039966	0.006058
BUTANE	3	6.83029	945.900	240.000	58.120	0.60200	0.001074	945.900	240.000			0.020295	0.007543
I-BUTANE	2	6.74809	882.800	240.000	58.120	0.58200	0.001240	882.800	240.000			0.062404	0.007441
BU-BENZ	2	6.98317	1577.960	201.378	134.210	0.87800	0.000845	1569.328	200.981			0.062404	0.007441
GBULACTO	2	11.01883	5426.398	463.250	86.090	1.13400	0.000360	1040.422	173.100			0.035662	0.007914
BUNITRIL		6.90200	1306.610	207.613	69.100	0.78920	0.000710	1216.298	379.487				
CETANE	2	7.03044	1831.320	154.527	226.430	0.78920	0.000700	4216.298	154.528				
CLPRNITR	2		1642.300	205.100	89.527	1.22400	0.001400	1871.381	150.046			0.058899	0.007128
CHRYSENE	2		2663.060	170.826	228.280	1.31200	0.000850	1992.040	214.136				
C-HEXANE	2	6.84498	1203.530	222.863	84.160	0.79863	0.000968	1203.530	222.863			0.154404	0.017638
C-C5ONE	2		958.490	214.300	84.110	0.92800	0.001000	1315.723	214.136			0.062404	0.007441
DECALIN	2		1501.270	154.540	138.240	0.88820	0.000810	958.490	154.540				
DECANE	3	6.95367	1480.000	194.480	142.280	0.75525	0.000760	1501.270	216.081			0.054551	0.007396
ET2-CO3	2		1951.000	215.100	106.160	1.14318	0.001110	966.509	95.127	3.0081	187.705	0.062404	0.007441
ET2-GLYC	2	8.25800	1281.200	118.200	86.130	0.99800	0.000472	1407.316	208.635	1.9631	1373.030	0.933050	-0.068586
MEZKTON	2	6.97840	1385.000	210.900	87.080	0.83200	0.000472	692.184	212.304				
MEZCNYH2			1329.100	206.700	60.064	0.95900	0.001000	1077.660	215.715				
DMF	2	6.99608	2693.420	208.724	73.094	0.96779	0.000950	1434.750	209.544			0.017907	0.006806
DMSO	2	8.41253	1994.000	133.200	78.110	1.00000	0.000677	1994.000	133.200			0.062404	0.007441
DIPHENYL	3		1437.840	197.900	282.536	0.40220	0.000680	1469.460	265.000			0.075749	0.014023
DODECANE	3	6.98059	1544.820	270.718	30.070	0.67100	0.000730	1531.330	146.254	-7.2049	-46.890	0.038008	0.008310
C12-BENZ	12	6.95719	1531.330	146.254	46.070	0.80641	0.000677	453.409	173.938			0.021631	0.009476
C12DECLN	3	6.99000	1652.050	231.480	106.160	0.88900	0.001043	1417.688	212.646			0.061015	0.009536
C12TETRL			1424.250	213.206	114.180	0.83610	0.000840	692.929	125.426	0.7301	125.427	0.316040	0.007441
C20H42	12		1347.500	180.600	80.520	1.23300	0.001000	1456.668	231.108			0.051421	0.019988
ETHANE	3	7.06640	1524.000	214.600	60.100	0.91600	0.000910	1335.780	209.570	-2.6136	8958.750	0.207883	-0.179450
ETHANOL	12	8.21337	1561.000	216.700	62.070	1.13694	0.001260	2258.460	198.395	9.9743	6102.828		
ET-BENZ	12	8.26210	2197.000	212.000	46.140	1.10151	0.001150	1909.803	206.296				
EBK	12		1893.400	243.200	100.080	1.03700	0.000860	1018.013	186.187			0.062404	0.007441
ECLHDRIN		7.55410	1200.560	224.210	100.200	0.70300	0.000635	1200.360	226.050			0.062404	0.007441
ETO-AMIN	12	6.79230	1268.110	216.900	100.200	0.70070	0.000850	268.110	216.900			0.047303	0.005296
ETG-BENZ	12	6.90240	2099.000	122.000	80.572	0.84010	0.000652	2099.100	122.000			0.062404	0.007441
E-OXLATE		7.07400	1135.410	226.572	86.170	0.67167	0.000926	1135.410	224.366			0.062404	0.007441
FURFURAL	12	6.83910	1171.530	224.366	86.170	0.67765	0.000917	1171.530	156.572			0.055066	0.015549
I-HEPTAN		6.87776	2081.000	170.000	212.320	0.96900	0.000733	1590.690	224.366				
N-HEPTAN	3	7.10390	1478.650	65.015	123.160	1.04650	0.001000	766.101	47.686				
C12-CHEX	2	7.21257	389.930	266.000	16.040	-0.23300	0.001110	389.930	266.000			0.062404	0.007441
I-HEXANE	3	6.61184	1574.990	238.800	32.050	0.81008	0.000940	1959.940	463.260	2.9242	463.529	0.497942	0.043529
N-HEXANE	2	8.07246	1503.600	215.400	76.000	0.98000	0.000950	1406.180	216.947				
C6NAPHTH														
33PIMPN														
METHANE														
METHANOL														
M-CELSOL														

NOTES --
IF COLUMN A IS BLANK, B AND C MAY BE REGRESSION CONSTANTS FROM ACTIVITY DATA
IF D AND E ARE BLANK, THERE WERE INSUFFICIENT SOLUTE DATA TO DETERMINE THEM
TYPE NUMBERS ARE 1 ASSOCIATING 2 POLAR 3 NON-POLAR
COMBINATIONS SUCH AS 12 POLAR AND ASSOCIATING ARE ALSO USED

Table 4.2 (*Continued*)

COMP.	TYPE	ANTOINE CONSTANTS			MOL. WT.	DENS. COEFFS.			E_A	T_L	Δs	Δh	d	e
		A_{VP}	B	C	M	a	b							
ME-CHEX	3	6.82889	1272.864	221.630	98.182	0.78675	0.000868		1272.864	221.630			0.062404	0.007441
MEK	12	6.38469	916.010	181.640	72.100	0.82570	0.001040		792.383	165.995	4.6506	2678.189	1.087090	-0.088043
ME-NAPH	2	7.06899	1852.670	197.716	142.190	1.02500	0.000436		1762.640	192.240			0.059129	0.008295
NM2PYRRO	2	8.02028	2371.240	259.220	99.130	1.02700	0.001000		1704.330	216.220				
NAPHTHAL	2	6.84577	1606.530	187.227	128.160	1.03040	0.000410		785.750	115.188				0.006987
NTROBENZ	2	7.08283	1772.200	199.000	123.108	1.22240	0.000960		1775.400	196.493			0.009269	
NTRIMETH	2	7.27417	1441.610	201.619	61.100	1.15570	0.001000		1018.116	197.055				
NONANE	3	6.93513	1428.810	201.619	128.250	0.73291	0.000764		1428.810	201.619			0.062404	0.007441
OCTADECA	3	7.04823	1920.600	144.530	254.480	0.79540	0.000680		1920.600	144.530			0.095066	0.007441
C18-BENZ	2	7.43600	2328.100	112.100	330.576	0.87160	0.000825		1721.650	96.579			0.047359	0.012620
C18-NAPH	2		1671.480	89.536	380.630	0.91950	0.000620		1536.415	83.207			0.062404	0.008313
2ME-HEPT	3	6.91735	1337.468	213.693	114.230	0.71392	0.000800		1337.468	213.693			0.062404	0.007441
I-OCTANE	3	6.81189	1257.840	220.735	114.220	0.70841	0.000824		1257.840	220.735			0.062404	0.007441
N-OCTANE	3	6.92377	1355.130	209.519	114.220	0.71864	0.000806		1355.130	209.519			0.062404	0.007441
I-PENTAN	3	6.78967	1020.010	233.097	72.150	0.63987	0.001010		1020.010	233.097			0.062404	0.007441
N-PENTAN	3	6.85221	1064.630	232.000	72.150	0.64771	0.000942		1064.630	232.000			0.062404	0.007441
1-PENTEN	3	6.84650	1044.890	233.516	70.130	0.66118	0.001034		1135.836	229.647			0.098426	0.015069
C5DIONE	2		1520.000	213.000	100.110	1.00000	0.001000		1400.650	211.613				
PHENANTH	2		512.660	91.605	178.220	1.11200	0.000890		108.972	117.839			0.016036	0.031996
PHENOL	12	9.54107	2744.400	247.624	94.110	0.50970	0.000610		822.636	243.182	1.4255	2006.330		
PROPANE	3	6.82973	813.200	248.000	44.090	0.79100	0.000360		813.200	248.000			0.062404	0.007441
PPNITRIL	2	6.92886	1285.780	220.000	55.080	0.79100	0.000360		1201.929	214.905				
PYRIDINE	2	7.73179	1774.000	250.805	79.100	1.00310	0.001010		695.596	349.907				
PYRROLID	2		1380.000	191.000	85.100	1.13600	0.000800		1441.815	211.571				
SULFOLAN	2	9.88629	4828.879	406.032	120.160	1.28670	0.000856		1466.055	172.863			0.085253	-0.008144
SO2	2	7.28228	999.900	273.190	64.070	1.43400	0.000700		659.538	140.803				
TERPHENL	2		1264.690	119.664	230.290	1.23400	0.000932		1264.690	119.664			0.037192	0.006711
N-C14H30	3	7.01245	1739.623	167.534	198.380	0.77707	0.000716		1739.623	167.534			0.062404	0.007441
TETRAHFN	2		1655.080	112.115	132.200	0.98800	0.000810		864.948	124.736			0.092094	0.015211
TOLUENE	2	6.95464	1344.800	219.482	92.130	0.88550	0.000928		1686.924	275.099			0.140559	0.084640
TRICONTA	2		2144.890	172.771	422.796	0.83200	0.001040		2141.610	102.771			0.063294	0.009347
N-TRIDEC	3	7.00339	1689.093	172.283	184.350	0.77022	0.000700		1689.093	174.283			0.062404	0.007441
TEG	12	8.35100	2078.600	90.003	150.170	1.14635	0.001070		966.325	65.949	-5.7981	452.744	-0.048717	
P-XYLENE	3	6.99052	1453.430	215.307	106.160	0.88245	0.001070		1453.430	215.307				0.006218

Table 4.3 Abbreviations Used in Table 4.2

COMPONENT NAME	ABBREVIATION
ACETONE	ACETONE
ACETONITRILE	AC-NITRL
ACETOPHENONE	AC-PHENO
ANILINE	ANILINE
ANTHRACENE	ANTHRAC
BENZENE	BENZENE
N-BUTANE	BUTANE
ISO-BUTANE	I-BUTANE
BUTYL BENZENE	BU-BENZ
GAMMA BUTYROLACTONE	GBULACTO
BUTYRONITRILE	BUNITRIL
CETANE	CETANE
BETA CHLOROPROPIONITRILE	CLPRNITR
CHRYSENE	CHRYSENE
CYCLOHEXANE	C-HEXANE
CYCLOPENTANONE	C-C5ONE
DECALIN	DECALIN
N-DECANE	DECANE
DIETHYL CARBONATE	ET2-CO3
DIETHYLENE GLYCOL	ET2-GLYC
DIETHYL KETONE	ET2-KTON
DIMETHYL ACETAMIDE	ME2ACNH2
DIMETHYL CYANAMIDE	ME2CYNH2
DIMETHYL FORMAMIDE	DMF
DIMETHYL SULFOXIDE	DMSO
DIPHENYL	DIPHENYL
DODECANE	DODECANE
DODECYLBENZENE	C12-BENZ
DODECYLDECALIN	C12DECLN
DODECYLTETRALIN	C12TETRL
EICOSANE	C20H42
ETHANE	ETHANE
ETHANOL	ETHANOL
ETHYL BENZENE	ET-BENZ
ETHYL BUTYL KETONE	EBK
ETHYLENE CHLOROHYDRIN	ECLHDRIN
ETHYLENE DIAMINE	E-DIAMIN
ETHYLENE GLYCOL	ETGLYCOL
ETHYL OXALATE	E-OXLATE
FURFURAL	FURFURAL
ISO-HEPTANE (2,2,3-TRIMETHYLBUTANE)	I-HEPTAN
N-HEPTANE	N-HEPTAN
HEXADECYL CYCLOHEXANE	C12-CHEX
ISO-HEXANE (2-METHYLPENTANE)	I-HEXANE
N-HEXANE	N-HEXANE
HEXYL NAPHTHALENE	C6NAPHTH
3,3PRIME-IMINODIPROPIONITRILE	33PIMPN
METHANE	METHANE
METHANOL	METHANOL
METHYL CELLOSOLVE	M-CELSOL

4.5 Summary

Table 4.3 (Continued)

COMPONENT NAME	ABBREVIATION
METHYL CYCLOHEXANE	ME-CHEX
METHYL ETHYL KETONE	MEK
METHYL NAPHTHALENE	ME-NAPH
N-METHYL-2-PYRROLIDONE	NM2PYRRO
NAPHTHALENE	NAPHTHAL
NITROBENZENE	NTROBENZ
NITROMETHANE	NTROMETH
NONANE	NONANE
OCTADECANE	OCTADECA
OCTADECYL BENZENE	C18-BENZ
OCTADECYL NAPHTHALENE	C18-NAPH
OCTANE (2-METHYLHEPTANE)	2ME-HEPT
ISO-OCTANE (2,2,4-TRIMETHYLPENTANE)	I-OCTANE
N-OCTANE	N-OCTANE
ISO-PENTANE (2-METHYLBUTANE)	I-PENTAN
N-PENTANE	N-PENTAN
1-PENTENE	1-PENTEN
PENTANEDIONE	C5DIONE
PHENANTHRENE	PHENANTH
PHENOL	PHENOL
PROPANE	PROPANE
PROPIONITRILE	PPNITRIL
PYRIDINE	PYRIDINE
PYRROLIDONE	PYRROLID
SULFOLANE	SULFOLAN
SULFUR DIOXIDE	SO2
TERPHENYL	TERPHENL
N-TETRADECANE	N-C14H30
TETRALIN	TETRALIN
TOLUENE	TOLUENE
TRICONTANE	TRICONTA
N-TRIDECANE	N-TRIDEC
TRIETHYLENE GLYCOL	TEG
P-XYLENE	P-XYLENE

4.5 SUMMARY

The important equations developed in this chapter are too numerous to list again. Activity coefficients for gas-phase solutions are usually assumed to be unity, although they can be calculated from an equation of state provided appropriate mixing rules are known to determine the constants for the equation of state for the mixture from the pure-component equation of state.

Activity coefficients for the liquid phase can be calculated from any one of a number of equations. Although these equations were usually derived by their advocates from theoretical considerations, they must be considered to be largely empirical since the constants must be obtained from experimental data. The constants of the equations are related, to some extent, to the infinite-dilution activity coefficient γ^∞. There are some theoretical and empirical correlations that can be used to estimate γ^∞; these correlations are most useful for systems exhibiting limited deviation from ideal-solution behavior, for screening purposes, or for calculations involving minor components that are not critical to product specifications.

The equations applicable to liquid-phase solutions are generally expected to apply to solid solutions as well.

REFERENCES

[1] O. Redlich and J. N. S. Kwong, *Chem. Rev.*, **44**, 233 (1949).
[2] J. M. Prausnitz and P. L. Chueh, *Computer Calculations for High Pressure Vapor-Liquid Equilibria*, Prentice-Hall, Englewood Cliffs, N.J., 1968.
[3] O. A. Hougen, K. M. Watson and R. A. Ragatz, *Chemical Process Principles*, 2nd ed., Part II, John Wiley & Sons, New York, 1959.
[4] G. N. Lewis and M. Randall, *Thermodynamics*, 2nd ed., revised by K. S. Pitzer and L. Brewer, McGraw-Hill Book Company, New York, 1961.
[5] H. C. Van Ness, *Classical Thermodynamics of Non-Electrolyte Solutions*, Macmillan, New York, 1964.
[6] J. H. Hildebrand and S. E. Wood, *J. Chem. Phys.*, **1**, 817 (1933).
[7] J. H. Hildebrand, *Proc. Natl. Acad. Sci. U.S.*, **13**, 267 (1927).
[8] J. H. Hildebrand, *J. Am. Chem. Soc.*, **51**, 66 (1929).
[9] J. H. Hildebrand and R. L. Scott, *Solubility of Non-Electrolytes*, 3rd ed., Reinhold, New York, 1950.
[10] J. H. Hildebrand and R. L. Scott, *Regular Solutions*, Prentice-Hall, Englewood Cliffs, N.J., 1962.
[11] G. Scatchard, *Chem. Rev.*, **8**, 321 (1931).
[12] M. Margules, *Sitzber. Akad. Wiss. Wien, Math. naturew.*, Klasse II, **104**, 1243 (1895).
[13] G. Scatchard and W. J. Hamer, *J. Am. Chem. Soc.*, **57**, 1805 (1935).
[14] J. J. Van Laar, *Z. Physik. Chem.*, **72**, 723 (1910).
[15] K. Wohl, *Trans. Am. Inst. Chem. Engrs.*, **42**, 215 (1946).
[16] G. M. Wilson, *J. Am. Chem. Soc.*, **86**, 127 (1964).
[17] P. J. Flory, *J. Chem. Phys.*, **9**, 660 (1941).
[18] M. L. Huggins, *Chem. Phys.*, **9**, 440 (1941).
[19] O. Redlich and A. T. Kister, *Ind. Eng. Chem.*, **40**, 341 (1948).
[20] C. Black, *Ind. Eng. Chem.*, **50**, 403 (1958).
[21] H. Renon and J. M. Prausnitz, *Am. Inst. Chem. Engrs. J.*, **14**, 135 (1968).
[22] K. C. Chao and J. D. Seader, *Am. Inst. Chem. Engrs. J.*, **7**, 598 (1961).
[23] R. F. Blanks and J. M. Prausnitz, *Ind. Eng. Chem. Fundamentals*, **3**, No. 2, 1 (1964).
[24] R. F. Weimer and J. M. Prausnitz, *Hydrocarbon Processing*, **44**, 237 (September 1965).
[25] J. G. Helpinstill and M. Van Winkle, *Ind. Eng. Chem. Proc. Des. Dev.*, **7**, 213 (April 1968).
[26] G. J. Pierotti, C. A. Deal, and E. L. Derr, *Ind. Eng. Chem.*, **51**, 95 (1959) and ADI Document No. 5782.
[27] H. R. Null and D. A. Palmer, *Chem. Engr. Progr.* **65**, 47 (September 1969).
[28] I. A. Wiehe and E. B. Bagley, *Ind. Eng. Chem. Fundamentals*, **6**, 209 (May 1967).
[29] C. Black, E. L. Derr and M. N. Papadopoulos, *Ind. Eng. Chem.*, **55**, 40 (1963).
[30] C. H. Deal and E. L. Derr, *Ind. Eng. Chem. Proc. Des. Dev.*, **3**, 394 (1964).
[31] J. A. Gerster, *J Chem. Eng. Data*, **5**, 423 (1960).

PROBLEMS

1. Derive an expression for the activity coefficients of a gaseous mixture by using Equation 4-12.

2. Derive an expression for the activity coefficient of a gaseous solution by using the virial equation truncated after the third term:

$$\frac{Pv}{RT} = 1 + \frac{B}{v} + \frac{C}{v^2},$$

where, for the mixture,

$$C = \sum_i \sum_j \sum_k x_i x_j x_k C_{ijk}.$$

3. Show that for a binary liquid system a Q-equation of the form $Q = Ax_1 + Bx_2$ implies an ideal solution.

4. Show that the Wohl equations (Equations 4-85 and 4-89) reduce to the binary forms (Equations 4-65 and 4-66) for two components.

5. Reduce the Fariss multicomponent equation (Equation 4-81) to the binary form.

6a. Estimate the three-suffix Van Laar constants for ethanol–isooctane at 50, 75, and 100°C.

6b. Calculate the values of the activity coefficients at mole-fraction increments of 0.1 for each of the three temperatures.

6c. Estimate the constants for the Wilson equations (Equations 4-68) for each of the three temperatures.

7. Repeat Problem 6 for ethanol–n-heptane and for n-heptane–isooctane.

8. Calculate, using the results of Problems 6 and 7, activity coefficients of an equimolar mixture of ethanol, n-heptane, and isooctane at 100°C. Use (a) the Wilson equation (Equations 4-68) and (b) Equation 4-105 with $N = 1$, 1.5, and 2.

Chapter 5

VAPOR–LIQUID EQUILIBRIUM

5.1 INTRODUCTION

The first four chapters have established the fundamentals involved in phase-equilibrium calculations. The remaining chapters deal with the practical problems that an engineer must solve.

This chapter deals with vapor–liquid equilibrium calculations. Processes taking advantage of favorable vapor–liquid distributions of components form by far the most extensive list of processes in which phase equilibrium is important. Distillation, the workhorse of the chemical and petroleum industries, is the classical example of an operation whose success is determined by a favorable distribution between vapor and liquid phases. There are, however, other operations of importance involving vapor–liquid equilibrium. Gas absorption, azeotropic distillation, extractive distillation, humidification and dehumidification, evaporation and flashing of liquid mixtures are all operations based on vapor–liquid equilibrium. In addition, many chemical reactions are carried out with a gas or vapor in intimate contact with a liquid phase. Even the simple operations of compression and expansion of gases and liquids often involve vapor–liquid equilibrium.

Because there are so many applications of vapor–liquid equilibrium, techniques for calculation and experimental determination of this particular type of phase equilibrium are more highly developed than for any other. It is therefore fitting that this subject should be the first of the applications to be mastered.

5.2 TYPES OF PHASE DIAGRAM IN VAPOR–LIQUID SYSTEMS

The different types of phase behavior occurring in vapor–liquid systems can be described most readily by referring to a binary system and to a quantity known as relative volatility. The relative volatility is defined in terms of an

5.2 Types of Phase Diagram in Vapor–Liquid Systems

equilibrium ratio, which is conventionally defined by

$$K_i \equiv \frac{y_i}{x_i}, \qquad (5\text{-}1)$$

where y_i is the mole fraction of component i in a vapor phase that is at equilibrium with a liquid phase containing x_i mole fraction of the same component. The relative volatility of component i with respect to j is conventionally defined by

$$\alpha_{ij} \equiv \frac{K_i}{K_j} = \frac{y_i \, x_j}{x_i \, y_j}. \qquad (5\text{-}2)$$

In a binary system $x_2 = 1 - x_1$ and $y_2 = 1 - y_1$. It is conventional in binary systems to choose as component 1 the component with the lower boiling point (the more volatile component), to express compositions in terms of component 1, and to drop subscripts, recognizing that $x = x_1$, $y = y_1$, and $\alpha = \alpha_{12}$. Thus the conventional binary equation is

$$\alpha = \frac{y(1-x)}{x(1-y)}, \qquad (5\text{-}3)$$

which, solved for y, becomes

$$y = \frac{\alpha x}{1 + (\alpha - 1)x}. \qquad (5\text{-}4)$$

The type of phase behavior exhibited by a binary system will be determined largely by the manner in which α varies with composition. For binary systems three methods are commonly used to represent the phase behavior graphically:

1. A plot of y versus x, which may be done either with temperature constant or with pressure constant.
2. A plot of total vapor pressure versus x (bubble-point curve) and of total vapor pressure versus y (dew-point curve) at constant temperature.
3. A plot of equilibrium temperature versus x and equilibrium temperature versus y at constant total pressure.

5.2.1 Ideal Solutions

The ideal solution has a constant value of α throughout all concentration ranges at constant temperature and a very nearly constant α at constant pressure. An example of the type of phase diagram exhibited by an ideal solution is shown in Figure 5.1. The y-x curve for such a system always lies above the $y = x$ line, which is usually shown on such diagrams, and exhibits

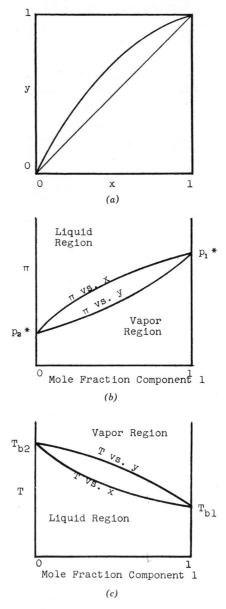

Figure 5.1 Phase diagrams for ideal solution: (*a*) isothermal *x-y* diagram; (*b*) isothermal π-composition diagram; (*c*) isobaric *T*-composition diagram.

5.2 Types of Phase Diagram in Vapor–Liquid Systems

no unusual behavior, such as maxima, minima, or discontinuities. Since the more volatile component is taken conventionally as component 1, α is always greater than unity. The isothermal π-composition curves are equally well behaved. The π-x curve is always above the π-y curve, with neither showing any peculiar behavior. The isobaric T-composition curves are similar, except that the T-y curve lies above the T-x curve. The benzene–toluene system is an example of an ideal solution.

5.2.2 Nonideal Systems, Nonazeotroping

Usually the value of α varies with composition. When its value is always greater than unity, the system is typically represented by Figure 5.2. In the system shown here α is greater at low values of x than at high values. Therefore the y-x curve has a tendency to inflect near a mole fraction of 1. The pressure-versus-composition and temperature-versus-composition curves also show this tendency to inflect, but there is no maximum nor minimum in any of the curves. In a system in which α is greatest at high values of x the inflection tendency is near $x = 0$. The region of the inflection tendency is frequently called a "pinch" to indicate that separations by distillation are difficult in this region. Systems of this type are quite common; methanol–water at atmospheric pressure is a typical example.

5.2.3 Low-Boiling Azeotropic System

If α varies greatly with composition, it might conceivably have values greater than unity at some concentrations and lower than unity at other concentrations. Since α is continuous in composition, it will equal unity at some concentration. If $\alpha = 1$, Equation 5-4 gives $y = x$ and the composition at which this occurs (i.e., vapor composition and liquid composition of a mixture are equal at equilibrium) is known as the azeotrope. If $\alpha > 1$ at low concentrations and $\alpha < 1$ at high concentrations, the azeotrope will be "low boiling." The term "low-boiling azeotrope" merely implies that, at a given pressure, the boiling temperature of the azeotrope is lower than the boiling temperature of either pure component.

Figure 5.3 shows a typical set of diagrams for a low-boiling azeotropic system. The y-x curve crosses the $y = x$ line at the azeotropic composition, the pressure-composition curves both exhibit maxima and coincide at the azeotropic composition, and for the isobaric system the temperature-versus-composition curves coincide at a minimum at the azeotropic composition. Such azeotropic systems frequently occur when the two components are dissimilar functionally and the boiling points are not greatly different (about 25°C for most systems, although much greater differences are noted in some

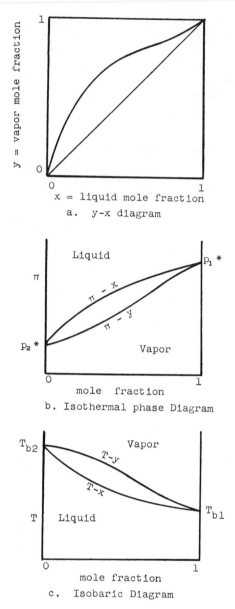

Figure 5.2 Phase diagrams for nonideal, nonazeotroping system: (*a*) *y-x* diagram; (*b*) isothermal phase diagram; (*c*) isobaric phase diagram.

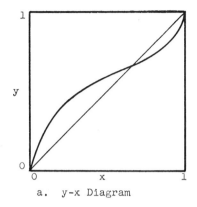

a. y-x Diagram

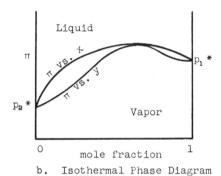

b. Isothermal Phase Diagram

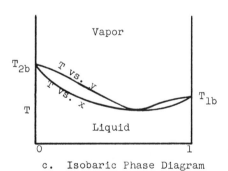

c. Isobaric Phase Diagram

Figure 5.3 Phase diagrams for low-boiling azeotropic system: (*a*) *y-x* diagram; (*b*) isothermal phase diagram; (*c*) isobaric phase diagram.

systems). Ethanol–water is a typical example. An extensive tabulation of known azeotropes is given by Horsley [1].

A very important characteristic of this type of system is that it is impossible by successive distillations at a given pressure (or at a given temperature) to obtain from a mixture both components as pure products. At some point in the procedure the azeotropic composition will be reached; when an azeotrope feed is partially vaporized, the vapor has the same composition as the liquid and no separation of components is effected.

5.2.4 High-Boiling Azeotropic Systems

If α varies with composition in such a way that it is less than unity at low concentration and greater than unity at high concentration, a high-boiling azeotrope occurs. This type of system differs from the low-boiling azeotrope in that the temperature-versus-composition curves of the isobaric plot exhibit *maxima* and the pressure-versus-composition curves of the isothermal plot exhibit *minima*. Typical diagrams for this type of system are shown in Figure 5.4. The high-boiling azeotrope is less common than the low-boiling azeotrope; it generally occurs between components whose molecules are somewhat attracted to each other. Acetone–chloroform is an example of this type of system.

5.2.5 Heterogeneous Azeotropic Systems

Frequently a system with a strong tendency to form a low-boiling azeotrope will consist of two components that are not completely miscible in the liquid phase. In such a system there will be a maximum solubility of component 1 in component 2, designated x_S, and a maximum solubility of component 2 in component 1, designated $1 - x'_S$. If a liquid mixture is prepared with overall composition between x_S and x'_S, it will separate into two layers, or phases. One phase will have composition x_S and the other x'_S; the relative quantities of the two liquid phases can be determined by material balance from the overall composition.

A typical set of diagrams for this type of system is shown in Figure 5.5. The compositions x_S and x'_S both have the same bubble point, and the equilibrium vapor composition y_a is the same for both liquid compositions.

A vapor having the azeotropic composition will, when condensed, form two liquid phases having mole fractions x_S and x'_S; thus the vapor having composition y_a is known as a heterogeneous azeotrope. Heterogeneous azeotropes are invariably low-boiling azeotropes, and this behavior is exhibited by most liquids with limited solubility. Since the liquid-phase composition is never equal to the vapor-phase composition, it is possible to

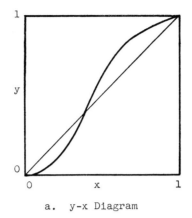

a. y-x Diagram

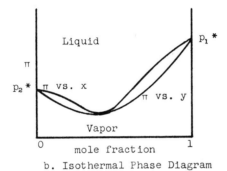

b. Isothermal Phase Diagram

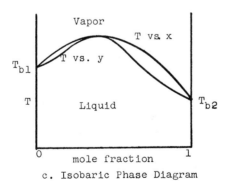

c. Isobaric Phase Diagram

Figure 5.4 Phase diagrams for high-boiling azeotropic systems: (*a*) *y-x* diagram; (*b*) isothermal phase diagram; (*c*) isobaric phase diagram.

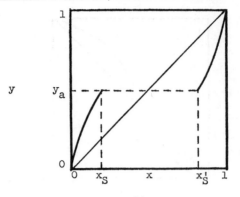

a. y-x Diagram

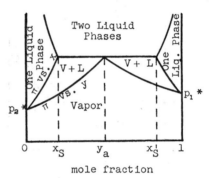

b. Isothermal Phase Diagram

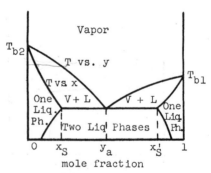

c. Isobaric Phase Diagram

Figure 5.5 Phase diagrams for heterogenous azeotroping system: (*a*) *y-x* diagram; (*b*) isothermal phase diagram; (*c*) isobaric phase diagram.

5.2 Types of Phase Diagram in Vapor–Liquid Systems

obtain both components pure by distilling a mixture; however, in a continuous distillation at least two columns are required, and in a batch distillation at least two runs are required. The n-butanol–water system is an example of this type of system.

5.2.6 Heterogeneous, Nonazeotroping System

It is possible for a partially miscible system not to exhibit an azeotrope. Such a system will have the equilibrium solubilities x_S and x'_S, and the same vapor composition will exist in equilibrium with both these liquid compositions. However, the equilibrium vapor composition will not lie between x_S and x'_S, as is the case with the heterogeneous azeotrope, and the boiling point of the x_S (or x'_S) composition lies between the boiling points of the pure components. Figure 5.6 shows the typical phase diagrams of such a system. Systems of this type are not numerous; an example is the propylene oxide–water system.

It is conceivable that a system might exhibit partial miscibility and have an azeotrope in the homogeneous region. The author is not aware of the actual existence of such systems.

5.2.7 Thermodynamic Interpretation

These various types of phase diagram can all be interpreted in terms of the thermodynamic equations developed in the preceding chapters. For the sake of simplicity in the qualitative discussion let us assume that the system being considered is a symmetrical system with an ideal gas phase and that the vapor pressures are of the same order of magnitude as the total pressure. Under these conditions Equation 3-13 applied to component 1, the more volatile one, becomes

$$\frac{y}{x} = \frac{\gamma_1 p_1^*}{\pi}. \tag{5-5}$$

Applied to component 2, Equation 3-13 gives

$$\frac{1-y}{1-x} = \frac{\gamma_2 p_2^*}{\pi}. \tag{5-6}$$

The definition of α, as given in Equation 5-3, yields

$$\alpha = \frac{\gamma_1 p_1^*}{\gamma_2 p_2^*}. \tag{5-7}$$

For an ideal solution $\gamma_1 = \gamma_2 = 1$ and

$$\alpha = \frac{p_1^*}{p_2^*}. \tag{5-8}$$

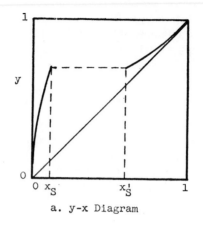

a. y-x Diagram

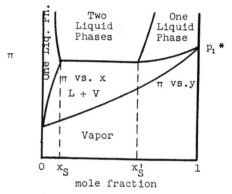

b. Isothermal Phase Diagram

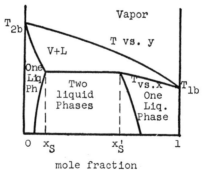

c. Isobaric Phase Diagram

Figure 5.6 Phase diagrams for a nonazeotroping, heterogeneous system: (*a*) y-x diagram; (*b*) isothermal phase diagram; (*c*) isobaric phase diagram.

5.2 Types of Phase Diagram in Vapor–Liquid Systems

It is this condition that gives rise to the simple type of phase diagram shown in Figure 5.1.

For the symmetrical system the liquid-phase activity coefficients are given by Equation (4-53):

$$\ln \gamma_1 = A(1-x)^2,$$
$$\ln \gamma_2 = Ax^2,$$

or

$$\frac{\gamma_1}{\gamma_2} = \exp[A(1-x)^2 - Ax^2]$$
$$= \exp[A(1-2x)]. \tag{5-9}$$

This gives, for α,

$$\alpha = \frac{p_1^*}{p_2^*} \exp[A(1-2x)]. \tag{5-10}$$

If the constant A is zero, we have an ideal solution. If A is positive, α varies from $(p_1^*/p_2^*)e^A$ at $x = 0$ to $(p_1^*/p_2^*)e^{-A}$ at $x = 1$. Since $e^A > e^{-A}$, α will be greater at low values of x than at high values of x and we shall have the type of diagram shown in Figure 5.2. In the event that A is negative α will be greater at high values of x and the inflection tendency will be nearer $x = 0$. These two conditions on A (or, more generally, on $\ln \gamma^\infty$) give rise to the terms "positive" and "negative" deviations from Raoult's law. (Raoult's law is the name given to the ideal-solution equation as applied to vapor–liquid equilibrium). These terms merely refer to the fact that $\ln \gamma^\infty$ is positive or negative, respectively. Thus for a positive deviation $\gamma^\infty > 1$ and for a negative deviation $\gamma^\infty < 1$.

Azeotropes occur whenever the value of α passes through unity. A typical plot of the variation of activity coefficient with composition is shown in Figure 5.7. At $x = 0$ the value of γ_1/γ_2 is simply γ_1^∞. The range of α, given by Equation 5-7, is

$$\gamma_1^\infty \frac{p_1^*}{p_2^*} < \alpha < \frac{1}{\gamma_2^\infty} \frac{p_1^*}{p_2^*},$$

or

$$\gamma_1^\infty \frac{p_1^*}{p_2^*} > \alpha > \frac{1}{\gamma_2^\infty} \frac{p_1^*}{p_2^*},$$

Thus, if α is to have a value of unity, the ratio p_2^*/p_1^* must lie between γ_1^∞ and $1/\gamma_2^\infty$. Hence the general criterion for an azeotrope to exist in a system in

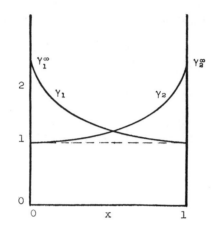

a. System With Positive Deviation

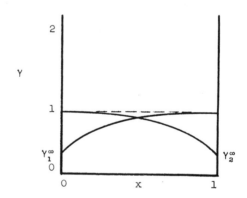

b. System with Negative Deviations

Figure 5.7 Typical composition dependency of activity coefficients in a symmetrical system: (*a*) system with positive deviation; (*b*) system with negative deviation.

5.2 Types of Phase Diagram in Vapor–Liquid Systems

which the gas phase is ideal is

$$\gamma_1^\infty < \frac{p_2^*}{p_1^*} < \frac{1}{\gamma_2^\infty}, \qquad (5\text{-}11a)$$

or

$$\gamma_1^\infty > \frac{p_2^*}{p_1^*} > \frac{1}{\gamma_2^\infty}. \qquad (5\text{-}11b)$$

If Equation 5-11a is true, a high-boiling azeotrope will exist, whereas a low-boiling azeotrope is associated with Equation 5-11b. This discussion indicates that there is no fundamental difference between systems that exhibit azeotropes and those that do not; the difference is merely one of the magnitude of the activity coefficients. Indeed, some systems exhibit azeotropes in one range of pressure (or temperature) and not in another.

Equation 3-31 was derived to indicate the conditions existing when two liquid phases are in equilibrium. For a binary system

$$\begin{aligned}\gamma_1 x_S &= \gamma_1' x_S', \\ \gamma_2(1 - x_S) &= \gamma_2'(1 - x_S').\end{aligned} \qquad (5\text{-}12)$$

If Equations 5-12 have any solution other than $x_S = x_S'$ in the region $0 \le x_S \le 1$, $0 \le x_S' \le 1$, the system will exhibit partial miscibility. The phase diagram will then be of the type shown in Figure 5.5 or 5.6 depending on whether the system is also azeotropic. In terms of the symmetrical system Equation 5-12 becomes

$$\begin{aligned}x_S \exp[A(1 - x_S)^2] &= \exp[A(1 - x_S')^2] x_S', \\ \exp(Ax_S^2)(1 - x_S) &= \exp(Ax_S'^2)(1 - x_S').\end{aligned} \qquad (5\text{-}13)$$

An examination of the symmetry of Equations 5-13 shows that the solution, if it exists, will give

$$x_S = 1 - x_S',$$

or

$$x_S' = 1 - x_S.$$

If the result is substituted in either of the equations, the result is

$$\exp[A(1 - x_S)^2] x_S = \exp(Ax_S^2)(1 - x_S). \qquad (5\text{-}14)$$

Equation 5-14 can be solved explicitly for A but not for x_S:

$$A = \frac{1}{1 - 2x_S} \ln \frac{1 - x_S}{x_S}. \qquad (5\text{-}15)$$

The solution to Equation 5-15 is shown graphically in Figure 5.8, which shows that if A is less than 2, there is no solution. Since $A = \ln \gamma^\infty$, the interpretation of this result is that, in a symmetrical system, if $\gamma^\infty > 7.39$, the system will exhibit partial liquid miscibility. If $\gamma^\infty \le 7.39$, the system is completely

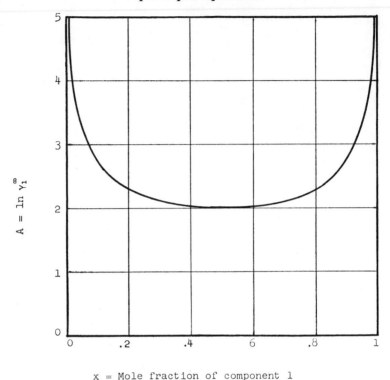

x = Mole fraction of component 1

Figure 5.8 Solubility limits of a symmetrical liquid system.

miscible at all concentrations. For nonsymmetrical systems there will be a locus of values of the parameters of the activity-coefficient equations dividing regions of total and partial miscibility.

The pertinent conclusion to be reached from the discussion in this section is that the occurrence of azeotropes and of partial or total miscibility is not indicative of gross basic differences between different systems; rather, these phenomena are indicative of the relative values of the activity coefficients of the components in the liquid phase.

5.2.8 High Pressures

There is qualitatively no difference in the types of phase diagram exhibited by systems with a nonideal gas phase and those discussed in the foregoing paragraphs, until the pressure of the system exceeds the critical value of one of the components. In the pressure range between the two critical pressures

5.2 Types of Phase Diagram in Vapor–Liquid Systems

the system exhibits a liquid phase over only a portion of the composition range, $x_A \leq x \leq 1$, where x_A is the minimum mole fraction at which a liquid phase exists. Above the critical pressure of the less volatile component some systems will exhibit a liquid phase over a composition range $x_A \leq x \leq x_B$, where $x_B < 1$.

Figure 5.9 shows the sequence of y-x and isobaric phase diagrams at increasing pressures beyond the critical point of each component. Curve I represents a pressure below either critical pressure; curve II is at the lower critical pressure; curve III represents a pressure between the two critical pressures; curve IV is at the higher critical pressure; and curve V is above both critical pressures. Systems at high pressures can also exhibit azeotropes and partial miscibility in the liquid phase. At such high pressures the densities of the gas and liquid phases are of the same order of magnitude and the gas-phase solution is just as likely to be nonideal as is the liquid phase.

In the region near the critical points of the components of a mixture some very peculiar behavior can be noted if one plots the pressure-versus-temperature locus of dew points and bubble points of a mixture of constant composition. For a single component the dew-point and bubble-point curves are identical, as illustrated by curve A of Figure 5.9c. The critical point (T_c, P_c) has three distinct characteristics:

1. All thermodynamic properties of the two phases at equilibrium are identical.
2. The temperature T_c is the maximum temperature at which a liquid and a vapor phase can exist at equilibrium.
3. The pressure P_c is the maximum pressure at which a liquid and a vapor phase can exist at equilibrium.

With a mixture, however, the bubble-point and dew-point curves are *not* identical. The three characteristics of the critical point could still describe a single point if the bubble-point and dew-point curves are similar to those designated for mixture B in Figure 5.9c. More often, however, the curves are similar to mixture C of Figure 5.9c. Characteristic 1 occurs at the intersection of the bubble-point curve with the dew-point curve. This point (T_c, P_c) is still considered the critical point of the mixture. It does not exhibit characteristics 2 and 3 of a pure component's critical point, however. The dew-point curve often passes through a maximum temperature; this point (T_{ct}, P_{ct}), is called the "cricondenherm"; the point of maximum pressure (T_{cp}, P_{cp}) is called the "cricondenbar."

A phenomenon known as retrograde condensation can occur in mixtures of type C. If the pressure is increased at constant temperature along dotted path I of Figure 5.9c, condensation will begin when the point α is reached. As pressure is further increased, there is more condensation; the condensate

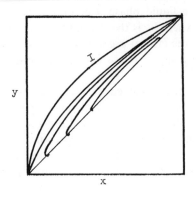

a. y-x Diagram

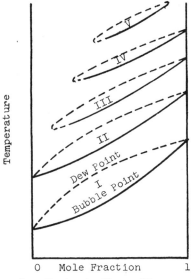

b. Isobaric Phase Diagram

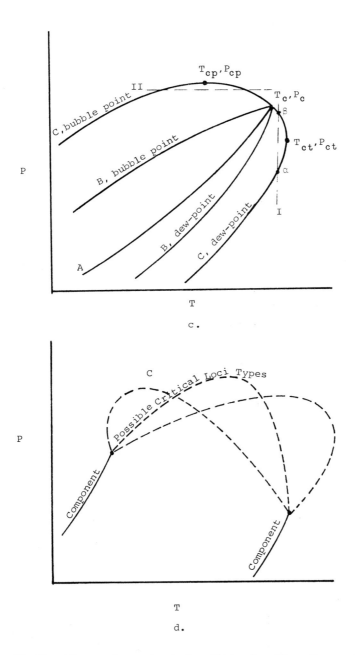

Figure 5.9 Phase diagrams for a system in the critical region: (a) y-x diagram; (b) isobaric phase diagram; (c) bubble-point and dew-point diagram; (d) critical-locus variations for mixtures.

will, of course, not have the composition of mixture C. As pressure is increased further, condensation will cease and the liquid will begin to vaporize. When point β is reached, the mixture is again completely vaporized. Thus we have a phenomenon of *vaporization with increasing pressure*. Similarly along dotted path II, when temperature is increased at constant pressure, we have first vaporization, followed by subsequent condensation. The critical temperatures and pressures of mixtures do not necessarily lie between the values of the pure-component critical points. Figure 5.9d shows three possible types of critical-point variation for mixtures. Kay [2] gives the critical locus, bubble-point, and dew-point curves for ethane–heptane.

5.2.9 Multicomponent Systems

The multicomponent system is very difficult to depict graphically. It is basically a combination of all the binary pairs, and each binary pair may exhibit any of the types of phase diagram already discussed. Many different

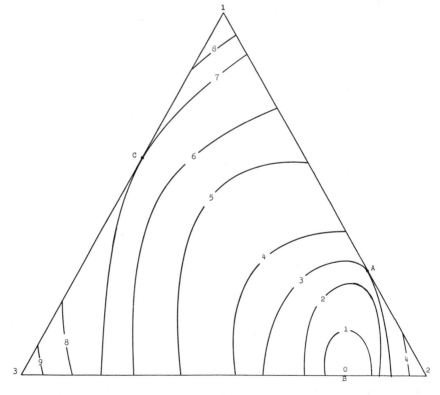

(a)

5.2 Types of Phase Diagram in Vapor–Liquid Systems

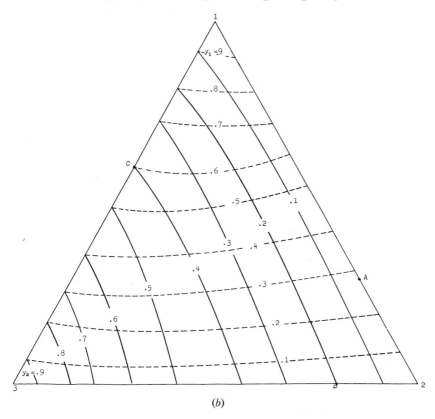

(b)

Figure 5.10 Ternary phase diagram of a system with three binary azeotropes but no ternary azeotrope: (a) lines of constant temperature versus liquid composition; (b) lines of constant vapor mole fraction versus liquid composition.

types of multidimensional surface are possible, with the degree of complexity increasing as the number of components increases. Figures 5.10, 5.11, and 5.12 show just three of the many possible triangular projections of an isobaric phase diagram of a ternary system. The system depicted in Figure 5.10 exhibits low-boiling azeotropes in each of the three binary pairs.

The triangular diagram represents the liquid-phase composition with one of the pure components at each apex. In plot a lines of constant bubble temperature are plotted on the composition triangle. Point A represents the azeotrope, or minimum boiling composition, of the 1-2 binary; point B, of the 2-3 binary; and point C, of the 1-3 binary. In this case point B represents the minimum boiling temperature of any mixture of the three components, and curves crossed proceeding away from B represent increasing temperature

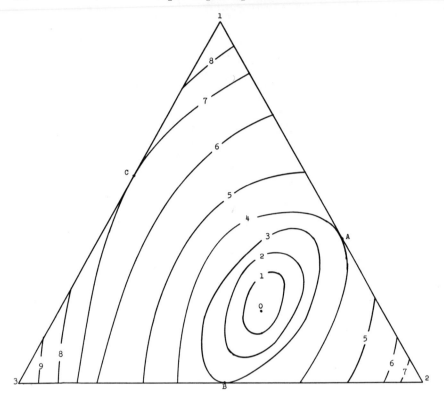

Figure 5.11 Diagram of a system with a ternary azeotrope. Lines of constant temperature versus liquid composition.

isotherms. The values shown on the isotherms are merely intended to represent relative values and do not represent any particular temperature scale. Plot *b* shows lines of constant values of y_1 and of y_2 on the liquid-composition triangle. Both plots are needed to represent completely the compositions and temperature of phases at equilibrium at constant pressure.

The system shown in Figure 5.11 exhibits a ternary azeotrope. Points *A*, *B*, and *C* represent the azeotropic compositions of the three binaries. In this system, however, there is a point, 0, whose boiling temperature is lower than any of the boiling temperatures of the three binary azeotropes. This composition is known as the ternary azeotrope. The lines of constant boiling temperature surround the azeotropic composition 0, and as one proceeds away from 0 the successive isotherms that are intersected represent increasing boiling temperatures. A plot showing lines of constant vapor composition would appear very much like Figure 5.10*a*. However, at the point represented

5.2 Types of Phase Diagram in Vapor–Liquid Systems

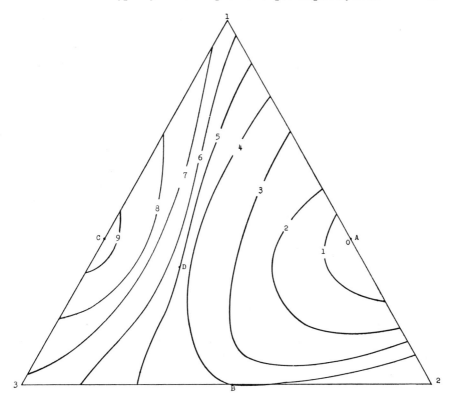

Figure 5.12 Ternary phase diagram exhibiting a saddle point.

by the composition of point 0 in Figure 5.11 the vapor composition would be identical with the liquid composition.

Figure 5.12 represents a system in which two of the binaries form low-boiling azeotropes, as indicated by points A and B. The third binary pair, however, forms a high-boiling azeotrope, represented by point C. If the plot were in the form of a three-dimensional surface, with temperature represented vertically on a triangular composition base, point C could be the highest point on the surface. Proceeding outward from C, the surface is of decreasing height, but convex. Somewhere along a line of steepest descent the surface will change from convex to concave. This is represented by point D, which is called the saddle point because of the similarity of the shape of the surface to a saddle.

The ternary phase diagrams become much more complex as additional complexities are involved in the binary pairs that make up the ternary. It is important to bear in mind, however, that these varied, complex diagrams result not from gross basic difference between the materials but rather from the relative differences in the magnitudes of the parameters in the activity-coefficient equations. Even the simplest of the activity-coefficient equations (other than ideal solution) is capable of producing all the types of phase diagrams discussed, and more.

5.3 EXPERIMENTAL VAPOR–LIQUID EQUILIBRIUM DATA

Although the process engineer is not often called on to obtain experimental vapor–liquid equilibrium data, he must often exercise judgment regarding the validity of the data and the relative costs and risks to be balanced between designing without additional experimental data desired and insisting that the data be obtained. In exercising such judgment it is advantageous for the engineer to be generally familiar with the types of laboratory apparatus and techniques, but he need not be familiar with all the details. The types of apparatus used may generally be classified as single-stage or multistage apparatus. The single-stage devices may be further classified into three types: static, recirculating, and flow. The multistage apparatus generally consists of a calibrated laboratory distillation column whose operation may be continuous or batch. In addition, chromatographic methods have been developed recently.

5.3.1 Single-Stage Vapor–Liquid Equilibrium Apparatus

An excellent discussion of all the important types of single-stage apparatus reported is given by Hala et al. [3]. This discussion is limited to the principles and problems involved.

Recirculating Stills. The typical recirculating still is shown schematically in Figure 5.13. The equilibrium between phases is established in the reboiler, which must be thoroughly agitated to give a uniform liquid composition. This agitation may be accomplished either by the boiling action of the vapor bubbles or by mechanical agitation. The vapor passes through a condenser, and the condensate is collected in a receiver. The condensate, which at steady state must have the same composition as the vapor, is returned through a condensate return line, which must be small enough to prevent backmixing between the reboiler and receiver. The return is usually accomplished by gravity flow, although pumping may be utilized.

The measurements consist of liquid composition (via a liquid sample from the reboiler), vapor composition (via a sample of the liquid condensate),

5.3 Experimental Vapor–Liquid Equilibrium Data

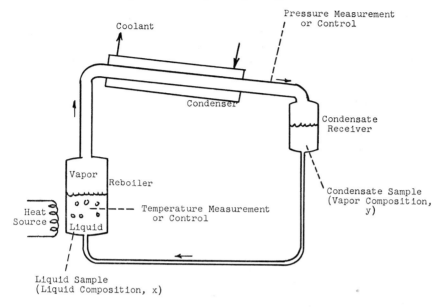

Figure 5.13 Typical schematic arrangement of a recirculating still.

reboiler temperature, and system pressure. If it is convenient to measure vapor composition directly from a vapor sample, the condenser and receiver may be replaced by a vapor pump.

The problems encountered are associated with ensuring that the measurements represent truly equilibrium conditions. In measuring temperature care must be taken that the temperature is that of vapor and liquid in equilibrium—not superheated liquid. The pressure drop between the reboiler and point of pressure measurement must be minimal.

The most severe problems, however, are associated with the composition measurements. Within the apparatus itself there must be absolutely no partial condensation of the vapor in the line between the liquid in the reboiler and the condenser. Such condensation would result in liquid reflux to the reboiler and a change in the vapor composition. Further care must be taken in the handling of the samples to ensure that none is lost through partial vaporization or condensation as the samples are removed from the pressure-and-temperature environment of the apparatus to that of the analysis. Any entrainment of liquid droplets into the condenser must also be avoided.

Measurements are generally made either isobarically, in which case pressure is controlled at a constant level and temperature measured, or isothermally, in which case temperature is controlled and pressure is measured. In making isothermal measurements the apparatus must be completely freed of inert gases, which would affect pressure measurements, and mechanical agitation is generally necessary since the rate of boiling (heat input) must be a part of the temperature-control loop. Data from recirculating stills are most frequently isobaric.

Static Apparatus. In a static apparatus a solution is charged into an evacuated chamber, which is immersed in a constant-temperature bath. After equilibrium is attained the pressure is measured and samples of vapor and liquid are withdrawn for analysis. This type of apparatus is shown schematically in Figure 5.14.

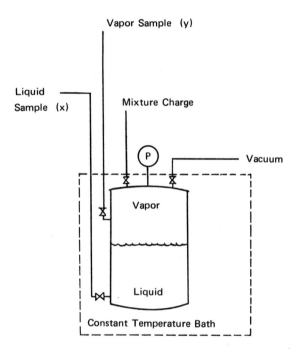

Figure 5.14 Static vapor-liquid equilibrium apparatus.

This method is deceptively simple in principle, but quite difficult to use in practice. Complete degassing is essential with each charge to ensure the accuracy of the pressure measurement. This degassing refers to dissolved gases in the liquid as well as evacuation of the apparatus before charging. Equilibration times are long, although they can be shortened by mechanical agitation or by pumping either liquid or vapor through the other phase

5.3 Experimental Vapor–Liquid Equilibrium Data

(i.e., allowing the liquid to spray in droplets through the vapor or the vapor to bubble through the liquid). Sampling in such a system can cause the equilibrium to change as the sample is withdrawn, although with modern instrumental analyses (particularly chromatography) large samples are not needed as frequently as they once were.

Flow Apparatus. The typical flow apparatus, of which there are many varieties, is shown schematically in Figure 5.15. One or more feed streams, which may be either vapor or liquid, are metered continuously into an equilibrium chamber where the total feed mixture is partially flashed. The liquid

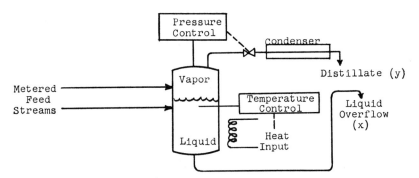

Figure 5.15 Flow vapor–liquid equilibrium apparatus.

effluent is withdrawn continuously by overflow and the vapor effluent by pressure control. Heat-input rate is controlled to give a set temperature in the reboiler. A change in either pressure or temperature changes the relative fraction of liquid and vapor output, and these pressure–temperature values must lie in a range between the bubble point and dew point of a mixture having the overall feed composition. The primary disadvantages are that the apparatus is inherently more complex and uses more material than other methods; its advantages are more rapid approach to equilibrium, relatively easier change of feed composition, and low residence time in the reboiler. This last advantage can be quite significant with heat-sensitive materials.

5.3.2 Multistage Apparatus

The multistage apparatus is usually a calibrated distillation column operating adiabatically at total reflux, as shown schematically in Figure 5.16. In practice a charge is placed in the reboiler, which is heated at a constant rate. Ultimately the vapors reach the condenser, where they are totally condensed, and all the condensate is returned to the top of the column. As liquid

94   *Vapor–Liquid Equilibrium*

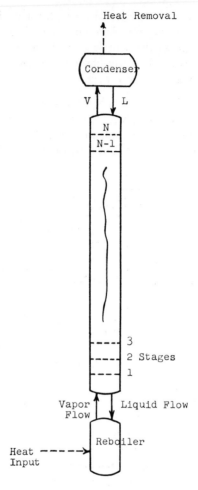

Figure 5.16   Total-reflux multistage-distillation apparatus.

builds up in the top stage, it ultimately overflows and drops to the next lower stage. Ultimately all stages are filled and overflowing, and at steady state the liquid downflow at each stage is equal to the vapor flowing upward, and the heat removed through the condenser is equal to that added at the reboiler.

At steady state a small sample is taken of the liquid entering the top of the column from the condenser and of that flowing to the reboiler from the bottom of the column. The temperature is read at these points also, and usually the overhead (condenser) pressure and pressure drop across the column are measured. The two compositions of the samples taken, of course, do not represent the composition of vapor and liquid at equilibrium; they

5.3 Experimental Vapor–Liquid Equilibrium Data

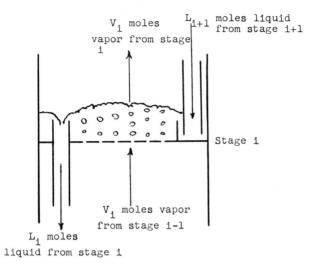

Figure 5.17 Schematic of a typical stage in a multistage vapor–liquid equilibrium apparatus.

represent the enrichment obtained after successive vapor–liquid contacting N times.

To analyze the results we must visualize an "ideal stage" in which the contact between liquid and vapor is sufficiently efficient for the vapor and liquid leaving the stage to be at equilibrium. The ideal stage is shown schematically in Figure 5.17. Since there is no feed to, nor product withdrawn from, the column, a material balance about the column above stage i will show that

$$V_i = L_{i+1},$$
$$y_{i,j} = x_{(i+1),j},$$

where i represents the stage number and j the component number. Similarly a material balance about the column below stage i will show that

$$V_{i-1} = L_i,$$
$$y_{(i-1),j} = x_{i,j}.$$

If the stage is ideal, $y_{i,j}$ and $x_{i,j}$ represent equilibrium compositions and the ratio of composition of two components is given by

$$\alpha_{jk} = \frac{y_{i,j} x_{i,k}}{x_{i,j} y_{i,k}} = \frac{x_{(i+1),j} x_{i,k}}{x_{(i+1),k} x_{i,j}}.$$

or

$$\frac{x_{(i+1),j}}{x_{(i+1),k}} = \alpha_{jk} \frac{x_{i,j}}{x_{i,k}}. \tag{5-16}$$

If a mean value of α_{jk} can be assumed throughout the column, Equation 5-16, applied successively, gives

$$\frac{x_{2,j}}{x_{2,k}} = \alpha_{jk} \frac{x_{1,j}}{x_{1,k}},$$

$$\frac{x_{3,j}}{x_{3,k}} = \alpha_{jk} \frac{x_{2,j}}{x_{2,k}} = \alpha_{jk}^2 \frac{x_{1,j}}{x_{1,k}},$$

$$\vdots$$

$$\frac{x_{(N+1),j}}{x_{(N+1),k}} = \alpha_{jk}^N \frac{x_{1,j}}{x_{1,k}}.$$

Solved for α_{jk}, this gives

$$\alpha_{jk} = \left[\frac{x_{(N+1),j}}{x_{(N+1),k}} \frac{x_{1,k}}{x_{1,j}} \right]^{1/N}. \qquad (5\text{-}17)$$

Solved for N, this gives

$$N = \frac{\log[(x_{(N+1),j}/x_{(N+1),k})(x_{1,k}/x_{1,j})]}{\log \alpha_{jk}}. \qquad (5\text{-}18)$$

Since $x_{(N+1),m}$ and $x_{1,m}$ ($m = 1, 2, \cdots$ number of components) represent the compositions of the two samples taken, the mean relative volatility for the compositions and conditions in the column can be calculated from Equation 5-17.

Since equilibrium is not usually attained on each stage, the number of ideal stages in the column, N, is not equal to the actual number of stages, N_A. The value of N must be obtained by making a run with a system whose α value is known and whose physical properties closely resemble those of the system to be determined. The value of N is then calculated from the experimental data and known α via Equation 5-18. Such a run is known as a calibration run.

After calibration the unknown system is run in the same column and at comparable operating conditions. The value of α_{jk} is then calculated from the data and known value of N via Equation 5-17.

The use of such multiple-stage procedures has two primary advantages:

1. If the value of α is near unity, the difference between vapor and liquid compositions can be small relative to experimental errors in a single-stage apparatus; the use of multistage contacting amplifies this difference to make analysis easier.

2. The column operation can be made to approximate conditions of plant operation more closely than any of the single-stage apparatuses.

The primary disadvantage is that only a single integral (or mean) value of α over the composition range and operating conditions existing in the actual column operation is obtained. Since the value of α is quite sensitive to composition in a nonideal solution, any extrapolation outside the range of the experimental values can be disastrously misleading. In particular, the existence of an azeotrope can never be correlated through the use of composition independent, integral α values.

It is, of course, possible to operate a multistage apparatus at conditions other than total reflux, but the mathematical analysis becomes much more complex.

The most popular type of multistage laboratory apparatus used for this purpose is a glass sieve-tray column known as an Oldershaw column. Actually, if it is adequately calibrated, any type of multistage-distillation column can be used for this purpose (packed column, wetted-wall column, etc.). The primary reason for use of the Oldershaw column is that column efficiency (N/N_A) is relatively insensitive to reboil rate between 30 and 90% of the column capacity and to small differences in the physical properties of the system, such as surface tension, viscosity, density, and relative volatility.

Further discussion of the use of distillation columns for measuring vapor–liquid equilibrium can be found in the periodical literature [4–8].

A different type of multistage method that is growing in popularity is the use of gas–liquid chromatography for measuring infinite-dilution activity coefficients [9–18].

A long column is filled with a solid support material, which is coated with the solvent. An inert carrier gas flows at a constant rate through the column. A small sample (about 1 μl) of liquid solute is injected into the gas stream and vaporized upstream of the column. The retention time (time lapse before the maximum concentration of the solute appears in the effluent from the column) is measured. The relative volatility of the solute is proportional to the difference between the retention time of the solute and residence time of the inert carrier gas. The activity coefficient of the solute in the solvent can be calculated from the relative volatility.

The primary difficulties with this method are the necessity for keeping the solute sample size low enough to justify the assumption of infinite dilution, the possibility of interaction between the solute and the solid support, and the necessity of operating conditions preventing evaporation of the solvent from the solid support.

5.4 PROCESSING VAPOR–LIQUID EQUILIBRIUM DATA

Experimental vapor–liquid equilibrium data must be interpreted quantitatively to yield the values of the pertinent parameters that allow the data to be used in process-design calculations. Usually the design calculation involves the calculation of vapor composition at equilibrium with a liquid of known composition, or vice versa. Equation 3-10 or 3-15 is the desired expression for this calculation:

$$\frac{y_i}{x_i} = \frac{(\gamma_i)_L}{(\gamma_i)_G} \frac{p_i^*}{\pi} \exp\left[\frac{(v_i)_L(\pi - p_i^*)}{RT}\right] \exp\left\{\frac{1}{RT}\int_{p_i^*}^{\pi}\left[\frac{RT}{P} - (v_i)_G\right] dP\right\}, \quad (3\text{-}10)$$

$$\frac{y_i}{x_i} = \frac{(\gamma_i)_L}{(\gamma_i)_G} \frac{f_i^*}{f_i^{\pi}} \exp\left[\frac{(v_i)_L(\pi - p_i^*)}{RT}\right]. \quad (3\text{-}15)$$

The only quantities in the above equations that are not pure-component physical properties are $(\gamma_i)_L$ and $(\gamma_i)_G$. Generally $(\gamma_i)_G$ is evaluated by one of three methods:

1. Assume ideal solution in the gas phase $[(\gamma_i)_G = 1]$.
2. Numerical or graphical integration of Equation 4-11,

$$RT \ln(\gamma_i)_G = \int_0^{\pi}(v_M - v_i) dP + \int_0^{\pi}\frac{\partial v_M}{\partial y_i} dP - \sum_j y_i \int_0^{\pi}\frac{\partial v_M}{\partial y_j} dP, \quad (4\text{-}11)$$

using pressure–volume–temperature measurements of gas mixtures.

3. Integration of Equation 4-11 or 4-41, using an equation of state with mixing rules for the parameters in the manner outlined in Section 4.2. If Equation 4-41 is used, the definition $\phi_i^{\pi} \equiv (\gamma_i)_G f_i^{\pi}$ is used.

With the gas-phase activity coefficients evaluated, the liquid-phase activity coefficient $(\gamma_i)_L$ remains. This quantity is generally obtained as a function of composition by correlation of the vapor–liquid equilibrium data via one of the equations presented in Sections 4.3.1, 4.3.2, and 4.3.3. In the absence of data one of the correlations for infinite-dilution activity coefficients outlined in Section 4.3.4 may be used, but their limitations must always be borne in mind.

There are actually four distinct problems involved in processing vapor–liquid equilibrium data, as follows:

1. Assessment of the quality of the data. (This is usually done by means of some thermodynamic consistency test.)
2. Choice of the functional form to represent the data. (One of the equations presented in Chapter 4 is usually sufficient.)
3. Choice of the variable to be used as a criterion for the statistical curve-fitting procedure.
4. Fitting the data to obtain the constants of the activity-coefficient equation.

The method of solving each of these problems depends somewhat on the type of data available. Several methods are discussed.

5.4 Processing Vapor–Liquid Equilibrium Data

5.4.1 Data on Pressure, Temperature, Vapor-Composition, and Liquid-Composition

When it is feasible to do so, the experimenter obtaining vapor–liquid equilibrium data should obtain and report total system pressure, temperature, liquid composition, and vapor composition at equilibrium. Although the system is completely defined even if one of these items is omitted, this data redundancy is quite useful in data analysis. The extent of trial and error in calculations is substantially reduced by having all these data available, and, in case one of the measurements is of marginal or unacceptable precision, that particular variable can be discarded from the data without negating the value of the remainder of the data. The computations involved in data analysis are much easier, for example, if the vapor composition is known experimentally; however, pressure, temperature, and liquid composition can often be measured with greater experimental accuracy (in cryogenic systems it is the liquid composition that presents the great difficulty).

Another advantage of redundancy in the data is that it offers a freedom of choice of objective function to be fitted. This is important since different applications often demand different criteria. The determination of the number of equilibrium stages in a column being designed is more accurate if the error in $\ln \alpha$ is minimized (see Equation 5-18).

Relative volatility is also the most significant parameter in determining reflux ratios, which are proportional to utility costs and required equipment capacity. On the other hand, if a precise knowledge of temperature or pressure is of more importance, one might wish to treat the data to minimize error in calculated temperature, pressure, activity coefficients, or equilibrium ratio (y/x). Such a knowledge of temperature or pressure is important in considering heat losses, corrosion rates, desired feed temperatures, and control systems based on temperature or pressure sensing as a measure of composition. The only criteria described in this work are minimization of error in calculated $\ln \alpha$, to be discussed in conjunction with complete P–T–x–y data, and minimization of fractional error in calculated total pressure, to be discussed with treatment of P–T–x data. The extension to other alternative criteria should pose no difficulty in principle.

Thermodynamic Consistency Tests. One of the greatest arguments in favor of obtaining redundant data is the ability to assess the validity of the data by means of a thermodynamic consistency test. All such tests are based on the application of Equation 2-15:

$$\sum_i x_i \, d\mu_i = -s \, dT + v \, dP. \tag{2-15}$$

It is usually more convenient to work with activity coefficients than with chemical potentials. Therefore Equation 2-18,

$$\mu_i = \mu_i^\circ + RT \ln \gamma_i x_i, \qquad (2\text{-}18)$$

is usually substituted into Equation 2-15 to yield

$$\sum_i x_i \, d\mu_i^\circ + RT \sum_i x_i \, d \ln \gamma_i + RT \sum_i x_i \, d \ln x_i$$
$$+ R \, dT \sum_i x_i \ln \gamma_i x_i = -s \, dT + v \, dP. \qquad (5\text{-}19)$$

By rearranging and making use of the relation $\sum_i x_i \, d \ln x_i = 0$, we can obtain

$$\sum_i x_i \, d \ln \gamma_i = -\frac{1}{T}\left(\frac{s}{R} + \sum_i x_i \ln \gamma_i x_i\right) dT + \frac{v \, dP}{RT} - \frac{1}{RT} \sum_i x_i \, d\mu_i^\circ. \qquad (5\text{-}20)$$

By making use of the fact that μ_i° is actually the molal free energy of the pure component i we obtain

$$d\mu_i^\circ = -s_i^\circ \, dT + v_i^\circ \, dP,$$

which, when incorporated into Equation 5-20, gives

$$\sum_i x_i \, d \ln \gamma_i = \frac{\Delta v^M}{RT} dP - \left(\frac{\Delta s^M}{R} + \sum_i x_i \ln \gamma_i x_i\right) \frac{dT}{T}, \qquad (5\text{-}21)$$

where Δv^M is the volume change due to mixing,

$$\Delta v^M = v - \sum_i x_i v_i^\circ,$$

and Δs^M is the entropy of mixing,

$$\Delta s^M = s - \sum_i x_i s_i^\circ.$$

It is also apparent that $RT \sum_i x_i \ln \gamma_i x_i$ is the free energy of mixing, since

$$RT \sum_i x_i \ln \gamma_i x_i = \sum_i x_i \mu_i - \sum_i x_i \mu_i^\circ$$

and $\sum_i x_i \mu_i = f$.

Thus

$$\sum_i x_i \, d \ln \gamma_i = \frac{\Delta v^M}{RT} dP - (T \Delta s^M + \Delta f^M) \frac{dT}{RT^2}. \qquad (5\text{-}22)$$

Since Δs^M and Δf^M represent property changes on mixing at constant temperature and pressure, $\Delta h^M = T \Delta s^M + \Delta f^M$, giving

$$\sum_i x_i \, d \ln \gamma_i = \frac{\Delta v^M}{RT} dP - \frac{\Delta h^M}{RT^2} dT. \qquad (5\text{-}23)$$

5.4 Processing Vapor–Liquid Equilibrium Data

An alternative form of Equation 5-21 is

$$\sum x_i \, d \ln \gamma_i = \frac{\Delta v^M}{RT} \, dP + \frac{\Delta h^M}{R} \, d\left(\frac{1}{T}\right). \tag{5-24}$$

Equation 5-24 gives a useful means of evaluating volume change on mixing and heat of mixing from activity coefficients:

$$\sum_i x_i \left(\frac{\partial \ln \gamma_i}{\partial P}\right)_{T, \text{all } x_j} = \frac{\Delta v^M}{RT}, \tag{5-25}$$

$$\sum_i x_i \left[\frac{\partial \ln \gamma_i}{\partial (1/T)}\right]_{P, \text{all } x_j} = \frac{\Delta h^M}{R}. \tag{5-26}$$

Equation 5-23 is normally applied to a binary system, in which case it becomes

$$x_1 \, d \ln \gamma_1 + x_2 \, d \ln \gamma_2 = \frac{\Delta v^M}{RT} \, dP - \frac{\Delta h^M}{RT^2} \, dT. \tag{5-27}$$

The values of γ_1 and γ_2 are obtained from the experimental data via Equation 3-10 or 3-15. Because of the inherent difficulty in obtaining accurate derivatives from experimental data, an integral form is usually applied:

$$\int_{x_1=a}^{x_1=b} x_1 \, d \ln \gamma_1 + \int_{x_1=a}^{x_1=b} x_2 \, d \ln \gamma_2 = \int_{x_1=a}^{x_1=b} \frac{\Delta v^M}{RT} \, dP - \int_{x_1=a}^{x_1=b} \frac{\Delta h^M}{RT^2} \, dT. \tag{5-28}$$

If the limits of integration are $x_1 = 0$ and $x_1 = 1$, a simplification can be made by integration by parts:

$$\int_{x_1=0}^{x_1=1} x_1 \, d \ln \gamma_1 = [x_1 \ln \gamma_1]_{x_1=0}^{x_1=1} - \int_0^1 \ln \gamma_1 \, dx_1$$

$$= -\int_0^1 \ln \gamma_1 \, dx_1$$

and

$$\int_{x_1=0}^{x_1=1} x_2 \, d \ln \gamma_2 = [x_2 \ln \gamma_2]_{x_2=1}^{x_2=0} - \int_{x_2=1}^{x_2=0} \ln \gamma_2 \, dx_2 = +\int_0^1 \ln \gamma_2 \, dx_1.$$

These two integrations lead to

$$\int_{x_1=0}^{x_1=1} x_1 \, d \ln \gamma_1 + \int_{x_1=0}^{x_1=1} x_2 \, d \ln \gamma_2 = \int_0^1 \ln \frac{\gamma_2}{\gamma_1} \, dx_1.$$

Thus the integral test over the range $0 \leq x_1 \leq 1$ becomes

$$\int_0^1 \ln \frac{\gamma_2}{\gamma_1} \, dx_1 = \int_{x_1=0}^{x_1=1} \frac{\Delta v^M}{RT} \, dP - \int_{x_1=0}^{x_1=1} \frac{\Delta h^M}{RT^2} \, dT. \tag{5-29}$$

The data utilized in thermodynamic consistency tests are either isobaric or isothermal. For isothermal data Equations 5-28 and 5-29 become

$$\int_{x_1=a}^{x_1=b} x_1 \, d \ln \gamma_1 + \int_{x_1=a}^{x_1=b} x_2 \, d \ln \gamma_2 = \int_{x_1=a}^{x_1=b} \frac{\Delta v^M}{RT} \, dP, \tag{5-30}$$

$$\int_0^1 \ln \frac{\gamma_2}{\gamma_1} \, dx_1 = \int_{x_1=0}^{x_1=1} \frac{\Delta v^M}{RT} \, dP. \tag{5-31}$$

For liquid mixing the quantity $\Delta v^M/RT$ is usually negligibly small and the right-hand side of the foregoing equations can be ignored, giving

$$\int_{x_1=a}^{x_1=b} x_1 \, d \ln \gamma_1 + \int_{x_1=a}^{x_1=b} x_2 \, d \ln \gamma_2 = 0, \tag{5-32}$$

$$\int_0^1 \ln \frac{\gamma_2}{\gamma_1} \, dx_1 = 0 \tag{5-33}$$

If the data are isobaric, the volumetric term drops out, giving, as the integral tests,

$$\int_{x_1=a}^{x_1=b} x_1 \, d \ln \gamma_1 + \int_{x_1=a}^{x_1=b} x_2 \, d \ln \gamma_2 = -\int_{x_1=a}^{x_1=b} \frac{\Delta h^M}{RT^2} \, dT, \tag{5-34}$$

$$\int_0^1 \ln \frac{\gamma_2}{\gamma_1} \, dx = -\int_{x_1=0}^{x_1=1} \frac{\Delta h^M}{RT^2} \, dT. \tag{5-35}$$

Unfortunately $\Delta h^M/RT^2$ is not, generally, a negligible quantity; thus the right-hand integral can be ignored only if the temperature range of the data is small, in which case Equations 5-32 and 5-33 would apply. Experimental values of Δh^M are not usually available either, although they can be estimated from Equation 5-26 if data are available at more than one pressure level and if the assumption

$$\left(\frac{\partial \ln \gamma_1}{\partial P} \right)_{T, \text{all } x_j} = 0$$

is valid. This latter assumption is equivalent to the assumption that $\Delta v^M/RT$ is negligible, as is usually assumed for the test of isothermal data.

The application of these tests can be visualized by following through the example problems that follow.

Example 5-1

Four sets of isothermal vapor–liquid equilibria for the ethanol–water system are given in Table 5.1. Figure 5.18 shows plots of the 25°C data of Dornte [19] and the 39.76°C data of Wrewski [20]. These two sets of data are not in agreement, and the higher temperature data of Wrewski [20], shown

Table 5.1 Isothermal Vapor–Liquid Equilibria for the Ethanol–Water System

$T = 25°C$[a]			$T = 39.76°C$[b]			$T = 54.81°C$[b]			$T = 74.79°C$[b]		
Mole % Ethanol in		Pressure (mm Hg)	Mole % Ethanol in		Pressure (mm Hg)	Mole % Ethanol in		Pressure (mm Hg)	Mole % Ethanol in		Pressure (mm Hg)
Liquid	Vapor		Liquid	Vapor		Liquid	Vapor		Liquid	Vapor	
12.2	47.4	41.8	0.00	0.00	54.3	0.00	0.00	116.6	0.00	0.00	286.9
16.3	53.1	45.2	6.89	45.6	81.4	9.16	47.53	192.9	8.95	45.64	469.2
22.6	56.2	47.9	8.03	47.30	84.7	11.57	50.36	204.2	13.17	50.83	508.5
32.0	58.2	50.7	9.94	49.23	90.5	21.20	57.27	228.1	14.20	51.94	513.7
33.7	58.9	51.0	14.52	54.31	99.4	23.75	58.28	233.9	15.40	52.68	519.7
43.7	62.0	52.7	15.48	55.16	101.2	26.71	58.88	237.3	19.20	55.30	539.5
44.0	61.9	52.8	18.31	57.19	104.4	36.98	61.51	247.5	26.28	57.68	566.6
57.9	68.5	54.8	22.08	58.74	107.3	47.88	65.54	256.6	36.48	60.54	590.0
83.0	84.9	58.4	23.33	58.76	108.6	61.02	71.02	264.6	49.13	65.32	615.9
			26.81	60.30	110.2	91.45	91.45	275.9	53.44	67.39	623.4
			36.77	63.41	115.7	100.0	100.0	275.2	64.78	72.90	640.5
			44.31	65.83	119.5				79.59	81.74	651.0
			48.08	67.26	121.9				89.65	89.65	654.0
			60.89	71.89	125.3				100.0	100.0	653.0
			77.96	81.29	129.2						
			93.90	93.97	131.5						
			95.52	95.52	131.4						
			100.0	100.0	129.8						

[a] Data from Dornte [19].
[b] Data from Wrewski [20].

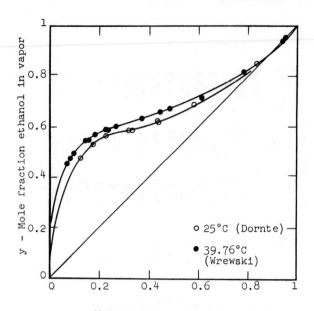

Figure 5.18 Vapor–liquid equilibria for ethanol–water system.

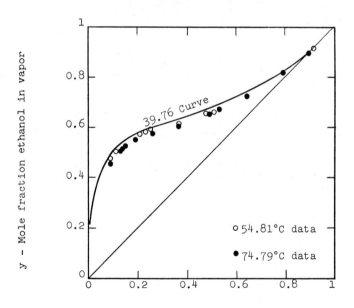

Figure 5.19 Vapor–liquid equilibria for ethanol–water system. Data from Wrewski [20].

5.4 Processing Vapor–Liquid Equilibrium Data

in Figure 5.19, would indicate that the 25°C curve should lie above the 39.76°C curve. Determine, from a thermodynamic consistency test, which of these sets of data is probably correct.

Solution. Since the data are isothermal, Equation 5-31 is used in the test. It has been asserted that the integral on the right-hand side of Equation 5-31 is usually negligible. It can be quickly verified that this is true. Lange's *Handbook of Chemistry* gives the following densities for ethanol–water solutions at 25°C:

Density (g/cc)	Ethanol (wt %)
0.99708	0
0.90985	50
0.78506	100

The values of the pure-component volumes of the two components are calculated from these data as follows:

$$v_1^\circ = \frac{\text{molecular weight}}{\text{density}} = \frac{46.07}{0.78506} = 58.71 \text{ cc/g-mole},$$

$$v_2^\circ = \frac{18.016}{0.99708} = 18.068 \text{ cc/g-mole}.$$

In order to calculate the molar volume of a 50% solution first calculate the molecular weight of the mixture:

$$\text{molecular weight} = \frac{1}{(0.5/18.016) + (0.5/46.07)} = 25.9,$$

$$v = \frac{25.90}{0.90985} = 28.47 \text{ cc/g-mole}.$$

The mole fraction x_1 is calculated from the weight fraction as follows:

$$x_1 = \frac{(0.5/46.07)}{(0.5/46.07) + (0.5/18.016)} = 0.2810.$$

The volume change due to mixing is calculated by application of the definition following Equation 5-21:

$$\Delta v^M = v - x_1 v_1^\circ - x_2 v_2^\circ = 28.47 - (0.281 \times 58.71)$$
$$- (0.719 \times 18.068) = 1.02 \text{ cc/g-mole}$$

In order to estimate the order of magnitude of the integral on the right-hand side of Equation 5-31 use the above value with the minimum temperature and total pressure range of this system as follows:

$$\int_{x_1=0}^{x_1=1} \frac{\Delta v^M}{RT} dP \text{ is on the order of } \frac{(-1.02)(58.8 - 23.7)}{(62{,}361)(298)} = -0.0000193.$$

This value, as will be seen in succeeding calculations, is truly negligible. Therefore Equation 5-33 is applicable. To apply the test we must plot the logarithm of γ_1/γ_2 versus x_1; consequently the values of γ_1 and γ_2 must be calculated for each of the data points. Since all pressures given in the data are subatmospheric, no appreciable error will be introduced if the vapor phase is assumed to be an ideal gas. Therefore Equation 3-11 can be used to calculate the activity coefficients. Furthermore, the magnitude of the correction term at 25°C can be estimated as follows:

$$\left| \exp\left[\frac{(v_i)_L}{RT}(\pi - p_1^*) \right] - 1 \right| \leq \exp\left[\frac{(58.71)(58.8 - 23.7)}{(62{,}361)(298)} \right] - 1 = 0.00111.$$

Since the correction factor will be very nearly unity, Equation 3-13 is applicable:

$$y_i = \frac{\gamma_i p_i^* x_i}{\pi},$$

Table 5.2 Calculated Results for Consistency Test of Ethanol–Water System

$T = 25°C$		$T = 39.76°C$		$T = 54.81°C$		$T = 74.79°C$	
x_1	$\ln \frac{\gamma_2}{\gamma_1}$	x_1	$\ln \frac{\gamma_2}{\gamma_1}$	x_1	$\ln \frac{\gamma_2}{\gamma_1}$	x_1	$\ln \frac{\gamma_2}{\gamma_1}$
0.122	−0.961	0.0689	−1.556	0.0916	−1.337	0.0895	−1.322
0.163	−0.852	0.0803	−1.459	0.1157	−1.189	0.1317	−1.097
0.226	−0.572	0.0994	−1.302	0.2120	−0.747	0.1420	−1.054
0.32	−0.176	0.1452	−1.074	0.2375	−0.642	0.1540	−0.988
0.337	−0.128	0.1548	−1.033	0.2671	−0.510	0.1920	−0.827
0.437	+0.166	0.1831	−0.914	0.3698	−0.143	0.2628	−0.519
0.440	0.425	0.2208	−0.743	0.4788	+0.131	0.3648	−0.160
0.579	0.450	0.2333	−0.672	0.6102	0.411	0.4913	+0.155
0.83	0.767	0.2681	−0.551	0.9145	0.859	0.5344	0.234
		0.3677	−0.220			0.6478	0.442
		0.4431	−0.013			0.7959	0.684
		0.4808	+0.075			0.8965	0.790
		0.6089	0.375				
		0.7796	0.666				
		0.9390	0.859				
		0.9552	0.871				

5.4 Processing Vapor–Liquid Equilibrium Data

or

$$\gamma_i = \frac{\pi y_i}{p_i^* x_i}.$$

This calculation is illustrated for the first data point of the 25°C data:

$$\gamma_1 = \frac{41.8 \times 0.474}{58.8 \times 0.122} = 2.762,$$

$$\gamma_2 = \frac{41.8 \times 0.526}{23.7 \times 0.878} = 1.0566,$$

$$\ln \frac{\gamma_2}{\gamma_1} = \ln \frac{1.0566}{2.762} = -0.961.$$

The results of all these calculations are tabulated in Table 5.2. Figure 5.20 shows a plot of the calculated data for the 25 and the 39.76°C data. Similar plots were made for all four sets of data and integrated graphically. The results are shown in Table 5.3.

Table 5.3 Results of Graphical Integrations for Example 5-1

$T(°C)$	$\int_0^1 \ln \frac{\gamma_2}{\gamma_1} dx$	$\int_0^1 \left| \ln \frac{\gamma_2}{\gamma_1} \right| dx$	Error (%)
25	0.0856	0.590	14.5
39.76	−0.0254	0.620	−4.1
54.81	−0.0299	0.653	−4.6
74.79	−0.0399	0.628	−6.4

The conclusion would be drawn from the values in Table 5.3 that the 25°C data of Dornte [19] are not thermodynamically consistent and the remainder of the data by Wrewski [20] are marginally acceptable. However, the data of Wrewski show a rather interesting phenomenon when the values of $\ln \gamma_1$ and $\ln \gamma_2$ are plotted versus mole fraction in the liquid, as shown in Figure 5.21. Since the logarithm of γ_2 must be 0 at $x_1 = 0$, the data indicate that $\ln \gamma_2$ must pass through a minimum somewhere in the neighborhood of $x_1 = 0.05$. Equation 5-27, applied to this situation, gives

$$x_1 \frac{d \ln \gamma_1}{dx_1} - x_2 \frac{d \ln \gamma_2}{dx_2} = 0.$$

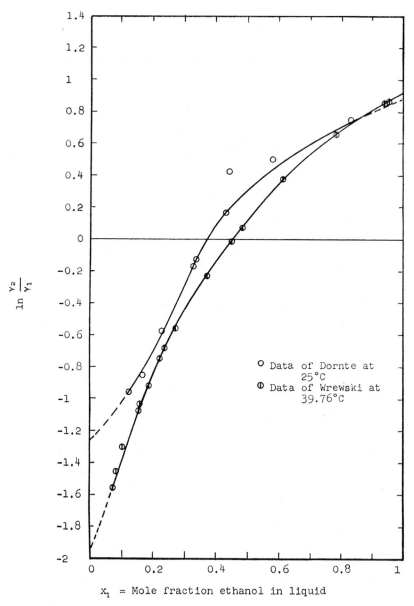

Figure 5.20 Thermodynamic consistency test of ethanol–water data.

5.4 Processing Vapor–Liquid Equilibrium Data

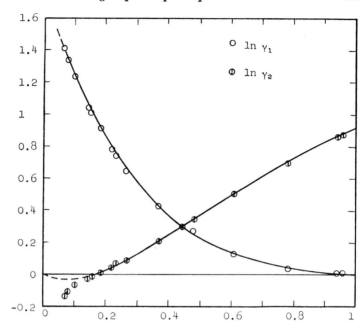

x_1 - Mole fraction ethanol in liquid

Figure 5.21 Activity coefficients calculated from experimental data exhibiting thermodynamic inconsistency.

Since $(d \ln \gamma_2/dx_1) = 0$ near $x_1 = 0.05$, $\ln \gamma_1$ should show a maximum at the same value. There is no indication in the data as plotted that such a maximum will occur. Consequently the conclusion to be drawn from the thermodynamic consistency tests of these data is that, although the data of Wrewski are apparently superior to those of Dornte, neither set is strictly consistent thermodynamically.

Example 5-2

Wilson and Simms [21] report isobaric vapor–liquid equilibrium data for the isopropanol–water system. Four sets of these data are shown in Table 5.4. Test these data for thermodynamic consistency.

Solution. Since the data are isobaric, Equation 5-35 is applicable. The activity coefficients are calculated in the same manner as indicated in Example 5-1. The results of these calculations are tabulated in Table 5.5. Figure 5.22 shows the plot of the atmospheric data, which must be used to evaluate the

Table 5.4 Isobaric Vapor–Liquid Equilibria for the Isopropanol–Water System

$P = 95$ mm Hg			$P = 190$ mm Hg			$P = 380$ mm Hg			$P = 760$ mm Hg		
Mole % Isopropanol in		T (°C)	Mole % Isopropanol in		T (°C)	Mole % Isopropanol in		T (°C)	Mole % Isopropanol in		T (°C)
Liquid	Vapor		Liquid	Vapor		Liquid	Vapor		Liquid	Vapor	
0.00	0.00	50.71	0.00	0.00	65.29	0.00	0.00	81.68	0.00	0.00	100.00
0.55	6.00	49.17	0.75	9.85	62.86	0.65	9.25	79.20	1.15	16.30	95.17
1.40	16.55	47.10	1.75	19.15	60.41	4.10	39.05	70.12	1.60	21.15	93.40
3.45	31.05	43.44	3.00	29.40	57.66	6.05	45.65	67.76	3.65	36.55	88.05
5.10	40.55	41.19	4.85	40.45	54.70	7.70	51.00	66.25	5.70	45.65	84.57
7.95	48.20	39.01	8.40	48.40	51.99	13.05	52.55	65.59	10.00	50.15	82.70
13.95	51.65	37.80	15.00	52.10	51.12	17.65	53.65	65.31	12.15	51.20	82.32
18.50	52.85	37.59	16.25	52.55	50.81	26.20	54.65	65.02	16.65	52.15	81.99
26.10	54.75	37.14	21.15	53.85	50.47	26.80	54.90	64.98	18.95	53.75	81.58

38.75	57.05	36.87	27.25	55.10	50.41	33.50	56.25	64.61	19.35	53.20	81.75
50.80	60.30	36.14	38.60	57.25	49.97	39.15	57.00	64.60	24.50	53.90	81.62
57.25	62.50	36.17	47.65	59.55	49.57	47.65	59.60	64.18	28.35	55.30	81.23
64.95	65.65	36.23	57.10	62.70	49.32	58.65	63.35	63.95	29.75	55.40	81.29
65.05	65.65	36.38	58.90	63.40	49.34	65.85	66.85	63.90	29.80	55.10	81.28
65.80	66.05	36.21	66.45	66.70	49.33	67.15	67.40	63.93	38.35	57.00	80.90
68.60	67.40	36.01	68.60	67.90	49.35	69.30	68.75	63.91	44.60	59.20	80.67
73.50	70.40	36.07	70.75	69.10	49.23	74.50	72.00	63.96	51.45	60.75	80.38
73.85	70.55	35.78	75.30	72.00	49.39	75.65	72.80	63.96	55.90	62.55	80.31
74.00	70.70	36.23	75.80	72.35	49.40	79.80	76.00	63.99	64.60	66.45	80.15
77.15	72.75	36.33	79.40	75.00	49.55	85.30	81.00	64.24	66.05	67.15	80.16
81.95	76.65	36.39	85.45	80.35	49.86	87.30	86.30	64.51	69.55	69.15	80.11
88.40	83.00	37.04	88.70	83.90	50.00	88.65	84.35	64.56	76.50	73.70	80.23
94.25	90.40	37.65	95.20	92.30	50.62	89.35	85.10	64.64	80.90	77.45	80.37
100.00	100.00	38.05	100.00	100.00	51.36	92.85	89.35	64.90	87.25	83.40	80.70
						100.00	100.00	66.02	95.35	93.25	81.48
									100.00	100.00	82.25

Table 5.5 Calculated Data for Thermodynamic Consistency Tests for the Isopropanol–Water System

	$P = 95$ mm Hg			$P = 190$ mm Hg			$P = 380$ mm Hg			$P = 760$ mm Hg	
x_1	$\ln \gamma_1$	$\ln \gamma_2$	x_1	$\ln \gamma_1$	$\ln \gamma_2$	x_1	$\ln \gamma_1$	$\ln \gamma_2$	x_1	$\ln \gamma_1$	$\ln \gamma_2$
0.0055	1.810	0.011	0.0075	2.028	0.013	0.0065	2.090	0.008	0.0115	2.152	0.005
0.0140	1.996	0.005	0.0175	1.959	0.026	0.0410	2.074	0.027	0.0160	2.149	0.015
0.0345	1.913	0.023	0.0300	1.979	0.032	0.0605	1.946	0.036	0.0365	2.076	0.022
0.0510	1.909	0.010	0.0485	1.960	0.021	0.0770	1.883	0.017	0.0570	1.990	0.025
0.0795	1.756	0.019	0.0840	1.723	0.047	0.1305	1.415	0.074	0.1000	1.597	0.059
0.1395	1.329	0.082	0.1500	1.260	0.091	0.1765	1.146	0.117	0.1215	1.438	0.078
0.1850	1.081	0.123	0.1625	1.205	0.111	0.2620	0.783	0.218	0.1665	1.155	0.124
0.2610	0.797	0.204	0.2115	0.982	0.161	0.2680	0.767	0.222	0.1895	1.073	0.134
0.3875	0.458	0.354	0.2725	0.755	0.217	0.3350	0.584	0.304	0.1935	1.034	0.144
0.5080	0.283	0.535	0.3860	0.467	0.359	0.3915	0.442	0.376	0.2450	0.817	0.200
0.5725	0.197	0.616	0.4765	0.316	0.483	0.4765	0.310	0.483	0.2835	0.712	0.237
0.6495	0.117	0.724	0.5710	0.199	0.613	0.5865	0.173	0.632	0.2975	0.664	0.252
0.6505	0.107	0.719	0.5890	0.178	0.636	0.6585	0.114	0.725	0.2980	0.657	0.260
0.6580	0.111	0.738	0.6645	0.109	0.745	0.6715	0.101	0.746	0.3835	0.454	0.362
0.6860	0.101	0.794	0.6860	0.094	0.774	0.6930	0.090	0.772	0.4460	0.350	0.426
0.7350	0.072	0.864	0.7075	0.087	0.812	0.7450	0.062	0.846	0.5145	0.245	0.531
0.7385	0.086	0.888	0.7530	0.057	0.875	0.7565	0.057	0.863	0.5590	0.194	0.583
0.7400	0.061	0.864	0.7580	0.055	0.883	0.7980	0.046	0.923	0.6460	0.117	0.699
0.7715	0.042	0.915	0.7940	0.037	0.935	0.8530	0.031	0.996	0.6605	0.104	0.720
0.8195	0.031	0.993	0.8545	0.017	1.027	0.8730	0.059	0.803	0.6955	0.084	0.768
0.8840	−0.002	1.082	0.8870	0.016	1.073	0.8865	0.019	1.047	0.7650	0.048	0.862
0.9425	−0.014	1.179	0.9520	0.009	1.161	0.8935	0.016	1.058	0.8090	0.036	0.910
						0.9285	0.015	1.108	0.8725	0.021	0.994
									0.9535	0.012	1.072

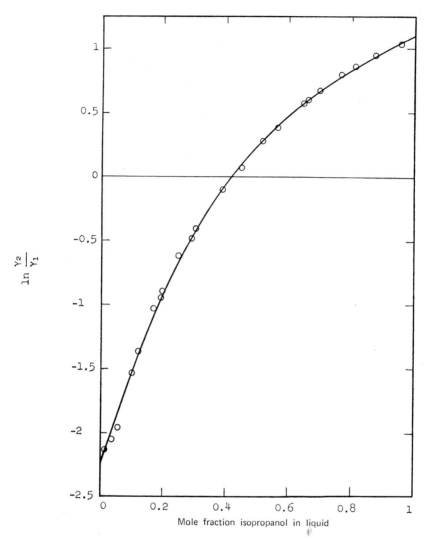

Figure 5.22 Thermodynamic consistency plot for isopropanol–water system at 760 mm Hg.

integral on the left side of Equation 5-25. The results of these graphical integrations are shown in Table 5.6.

Table 5.6   Results of Graphical Integrations for Example 5-2

Pressure (mm Hg)	$\int_0^1 \ln \frac{\gamma_2}{\gamma_1} dx$	$\int_0^1 \left| \ln \frac{\gamma_2}{\gamma_1} \right| dx$	Departure from Zero (%)
95	+0.025	0.817	3.06
190	0	0.79	0
380	−0.01175	0.794	1.48
760	−0.0185	0.774	2.39

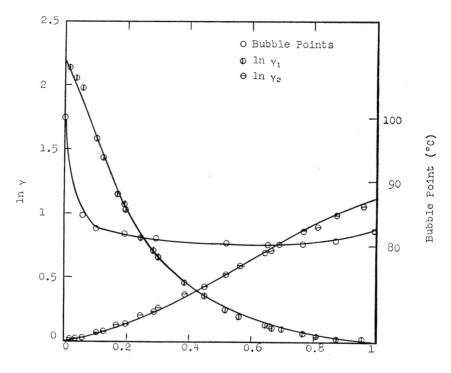

Figure 5.23 Temperatures and activity coefficients for isopropanol–water system at 760 mm Hg.

5.4 Processing Vapor–Liquid Equilibrium Data

In order to complete the test the integral on the right-hand side of Equation 5-35 must be estimated. This is done from the plot of the logarithm of the activity coefficients and temperature versus x_1 shown in Figure 5.23. Similar plots were prepared for each set of data, and values of temperature and the logarithm of the activity coefficients at constant values of x_1 were read from the plots. These values are shown in Table 5.7. The heat of mixing is evaluated by means of Equation 5-26 from plots of $\ln \gamma_1$ and $\ln \gamma_2$ versus $1/T$. Figure 5.24

Table 5.7 Values of $\ln \gamma$ versus Temperature at Constant Mole Fraction (Isopropanol–Water System)

Mole Fraction Isopropanol in Liquid	$T(°C)$	$\ln \gamma_1$	$\ln \gamma_2$
0.05	85.7	1.96	0.02
	70.0	2.05	0.025
	54.6	1.98	0.02
	41.5	1.98	0.02
0.1	82.8	1.62	0.05
	66.5	1.62	0.05
	52.5	1.59	0.05
	38.5	1.51	0.05
0.2	81.7	1.00	0.14
	65.0	1.04	0.13
	50.6	1.03	0.13
	37.4	1.04	0.13
0.5	80.4	0.28	0.5
	63.6	0.28	0.5
	49.4	0.27	0.52
	36.5	0.29	0.51
0.8	80.4	0.05	0.88
	61.8	0.03	0.92
	49.5	0.05	0.89
	36.3	0.04	0.97
0.9	81.0	0.01	1.00
	64.5	0.0	1.05
	50.3	0.01	1.08
	37.0	0.01	1.17
0.95	81.6	0.0	1.06
	65.0	0.0	1.12
	50.9	0.0	1.14
	37.5	0.0	1.27

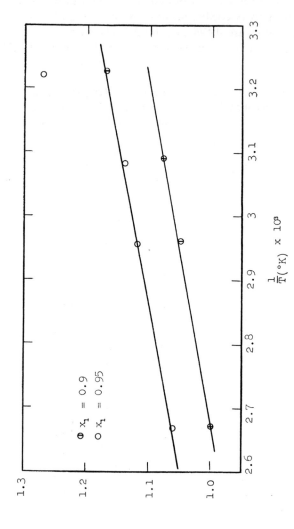

Figure 5.24 Plots for determining the heat of mixing from activity coefficients.

5.4 Processing Vapor–Liquid Equilibrium Data

shows such a plot for component 2 at two values of x_1. The slopes of the lines in Figure 5.24 were determined, and from these values the value of the partial molal heat of mixing was evaluated as follows:

At $x_1 = 0.90$,

$$\frac{\partial \ln \gamma_2}{\partial (1/T)} = \frac{(1000)(1.079 - 0.985)}{(3.2 - 2.7)} = 188,$$

$$\frac{\partial \ln \gamma_1}{\partial (1/T)} = 0,^\dagger$$

$$\frac{\Delta h^M}{1.987} = 0.1 \times 188 = 18.8,$$

$$\Delta h^M = 37.4 \text{ cal/g-mole}.$$

An approximate value for the integral was obtained by using 37.4 cal/g-mole as an average value of the heat of mixing, the total temperature range, and the minimum temperature throughout the range as follows:

$$\int_{x_1=0}^{x_1=1} \frac{\Delta h^M}{RT^2} dT \simeq \frac{(37.4)(20)}{(2)(350)^2} = 0.00305.$$

Since this value is near zero as compared with the total areas of the curves involved, the conclusion can be drawn that these data are thermodynamically consistent. If the temperature range had been larger, the integral would not have been negligible and a more careful evaluation of the integral would have been necessary. The reader is cautioned *not* to conclude from this example that

$$\int_{x_1=0}^{x_1=1} \frac{\Delta h^M}{RT^2} dT$$

is generally negligible.

Data Analysis Based on Minimal Error in Relative Volatility. In the past a number of graphical means have been developed to determine the constants of various vapor–liquid equilibrium equations from the experimental data. These have generally depended on the specific type of equation used, and no generalization can be made. Some of the graphical procedures used on binary systems are outlined by Reid and Sherwood [22].

With the increasing availability of digital computers, however, it is possible to develop general methods that do not depend on the specific activity-coefficient equation used. When experimental vapor–liquid equilibrium data are available in terms of compositions of both vapor and liquid phases, it is possible to correlate the data by least-squares minimization of the difference

† See Table 5.7.

between calculated and experimental values of the logarithm of the relative volatility. Such a basis is desirable when one wishes to use the data to determine the number of stages of distillation required to effect a given separation. Equation 5-18 shows the direct relationship between the number of stages required and the logarithm of the relative volatility for a system at total reflux.

The experimental value of the relative volatility can be determined by application of the definition of relative volatility as given in Equation 5-2, repeated below.

$$\alpha_{\text{exp}} = \frac{y_1 x_2}{y_2 x_1}. \tag{5-2}$$

A calculated value of the relative volatility can be obtained by calculating the y/x ratio for each of the components, as indicated below:

$$\alpha_{\text{calc}} = \frac{(y_1/x_1)_{\text{calc}}}{(y_2/x_2)_{\text{calc}}}. \tag{5-36}$$

The y/x ratios are calculated by application of either Equation 3-10 or Equation 3-15:

$$\left(\frac{y_1}{x_1}\right)_{\text{calc}} = \frac{(\gamma_1)_L}{(\gamma_1)_G} \frac{p_1^*}{\pi} \exp\left[\frac{(v_1)_L(\pi - p_1^*)}{RT}\right] \exp\left\{\frac{1}{RT}\int_{p_1^*}^{\pi}\left[\frac{RT}{P} - (v_1)_G\right]dP\right\},$$

$$\left(\frac{y_2}{x_2}\right)_{\text{calc}} = \frac{(\gamma_2)_L}{(\gamma_2)_G} \frac{p_2^*}{\pi} \exp\left[\frac{(v_2)_L(\pi - p_2^*)}{RT}\right] \exp\left\{\frac{1}{RT}\int_{p_2^*}^{\pi}\left[\frac{RT}{P} - (v_2)_G\right]dP\right\}, \tag{3-10}$$

$$\left(\frac{y_1}{x_1}\right)_{\text{calc}} = \frac{(\gamma_1)_L}{(\gamma_1)_G} \frac{f_1^*}{f_1^\pi} \exp\left[\frac{(v_1)_L(\pi - p_1^*)}{RT}\right],$$

$$\left(\frac{y_2}{x_2}\right)_{\text{calc}} = \frac{(\gamma_2)_L}{(\gamma_2)_G} \frac{f_2^*}{f_2^\pi} \exp\left[\frac{(v_2)_L(\pi - p_2^*)}{RT}\right]. \tag{3-15}$$

Equations 3-10 or 3-15, substituted into Equation 5-36, gives for the calculated relative volatility

$$\alpha_{\text{calc}} = \frac{(\gamma_1)_L(\gamma_2)_G p_1^*}{(\gamma_2)_L(\gamma_1)_G p_2^*} \frac{\exp\left[\frac{(v_1)_L(\pi - p_1^*)}{RT}\right] \exp\left\{\frac{1}{RT}\int_{p_1^*}^{\pi}\left[\frac{RT}{P} - (v_1)_G\right]dP\right\}}{\exp\left[\frac{(v_2)_L(\pi - p_2^*)}{RT}\right] \exp\left\{\frac{1}{RT}\int_{p_2^*}^{\pi}\left[\frac{RT}{P} - (v_2)_G\right]dP\right\}}, \tag{5-37}$$

$$\alpha_{\text{calc}} = \frac{(\gamma_1)_L(\gamma_2)_G f_1^* f_2^\pi}{(\gamma_2)_L(\gamma_2)_G f_2^* f_1^\pi} \frac{\exp\left[\frac{(v_1)_L(\pi - p_1^*)}{RT}\right]}{\exp\left[\frac{(v_2)_L(\pi - p_2^*)}{RT}\right]}. \tag{5-38}$$

5.4 Processing Vapor–Liquid Equilibrium Data

The fugacities, volumes, and vapor pressures in the above equations are pure-component properties and can be determined from the thermodynamic properties of the pure components. The activity coefficients are mixture properties and must be calculated by means of the type of equations outlined in Chapter 4. When the activity-coefficient equation to be used has been chosen, α_{calc} becomes a function of the parameters that are used in the activity-coefficient equation, designated by A_m. The parameters A_m are chosen to minimize the value of the variable $\mathscr{S}$:

$$\mathscr{S} = \sum_{k=1}^{L} R_k^2, \tag{5-39}$$

where L is the total number of data points. The term R_k is defined by

$$R_k = (\ln \alpha_{\text{exp}})_k - (\ln \alpha_{\text{calc}})_k. \tag{5-40}$$

In order for the summation $\mathscr{S}$ to have its minimum value, the values of all its partial derivatives with respect to the parameters A_m must be zero:

$$\frac{\partial \mathscr{S}}{\partial A_m} = 0 \quad \text{(for all } m\text{).} \tag{5-41}$$

The derivatives of the residuals are in turn related to the calculated value of the relative volatility:

$$-\frac{\partial R_k}{\partial A_m} = \frac{\partial (\ln \alpha_{\text{calc}})_k}{\partial A_m}. \tag{5-42}$$

Thus

$$\frac{\partial \mathscr{S}}{\partial A_m} = -2 \sum_{k=1}^{L} \left(\ln \frac{\alpha_{\text{exp}}}{\alpha_{\text{calc}}} \right)_k \left(\frac{\partial \ln \alpha_{\text{calc}}}{\partial A_m} \right). \tag{5-43}$$

Thus it is necessary to find a set of values of A_m that makes the partial derivatives of $\mathscr{S}$ as calculated by Equation 5-43 all equal to zero. A trial-and-error computational method must be used, and it is usually futile to attempt to do the calculation without the aid of a digital computer.

Although a thorough discussion of search procedures for minimizing a function is beyond the scope of this work, a brief description of two of the more commonly used search methods is presented. The first of these methods is known as steepest descent. With this procedure, the parameters A_m are changed simultaneously in such a manner that the change in each A_m is proportional to the negative of the partial derivative of $\mathscr{S}$ with respect to that A_m. A search is made in the direction thus indicated until the minimum value of $\mathscr{S}$ along this line is located. At this minimum point the derivatives are again evaluated and another search procedure is initiated in the direction indicated at the new point. The search procedure is repeated until the values of all the partial derivatives are sufficiently near zero. This method is quite efficient at values of A_m relatively far away from the optimum set, but as the optimum values are approached the step size for each iteration becomes quite small and a very large number of iterations is required.

A second frequently used search procedure is known as the Newton–Raphson method. The change of each partial derivative associated with changes in the parameters is estimated by the following set of equations:

$$\Delta\left(\frac{\partial \mathscr{S}}{\partial A_1}\right) = \frac{\partial^2 \mathscr{S}}{\partial A_1^2} \Delta A_1 + \frac{\partial^2 \mathscr{S}}{\partial A_1 \partial A_2} \Delta A_2 + \cdots,$$

$$\Delta\left(\frac{\partial \mathscr{S}}{\partial A_2}\right) = \frac{\partial^2 \mathscr{S}}{\partial A_1 \partial A_2} \Delta A_1 + \frac{\partial^2 \mathscr{S}}{\partial A_2^2} \Delta A_2 + \cdots,$$

$$\vdots \qquad \vdots \qquad \vdots$$

(5-44)

Since the object of the search is to find values of the parameters for which the partial derivatives are zero, the change in each partial derivative is set equal to minus the partial derivative itself:

$$\Delta\left(\frac{\partial \mathscr{S}}{\partial A_m}\right) = -\frac{\partial \mathscr{S}}{\partial A_m}.$$

Thus

$$-\frac{\partial \mathscr{S}}{\partial A_1} = \frac{\partial^2 \mathscr{S}}{\partial A_1^2} \Delta A_1 + \frac{\partial^2 \mathscr{S}}{\partial A_1 \partial A_2} \Delta A_2 + \cdots,$$

$$-\frac{\partial \mathscr{S}}{\partial A_2} = \frac{\partial^2 \mathscr{S}}{\partial A_1 \partial A_2} \Delta A_1 + \frac{\partial^2 \mathscr{S}}{\partial A_2^2} \Delta A_2 + \cdots.$$

(5-45)

When the set of Equations 5-45 has been solved for values of ΔA_1, ΔA_2, etc., new values of the parameters A_m are then calculated by

$$(A_m)_{\text{new}} = (A_m)_{\text{old}} + \lambda \, \Delta A_m.$$

(5-46)

The proper value of λ is found by a trial-and-error procedure to give the minimum value of $\mathscr{S}$ in the indicated direction. The Newton–Raphson search method is quite efficient in the neighborhood of the optimum set of parameters, but at points far removed from the optimum it is quite inefficient and can even be nonconvergent. It can seek a maximum value as well as a minimum value and is incapable of distinguishing which it is searching. Most practical search methods make use of both of the methods described above, incorporating some criterion for determining which method to use at each point in the search. One of the most powerful techniques for such searches is described by Fariss [23].

5.4 Processing Vapor–Liquid Equilibrium Data

The data for a large number of systems have been fitted to various activity-coefficient equations to determine the applicability of the equations, using least-squares fit to ln α as a criterion. The results obtained with two-constant equations and isothermal data are shown in Table 5.8. The gas phase was considered to be an ideal solution in these tests, but the pure components were corrected for deviations from the ideal gas law according to the Redlich–Kwong equation. The tables show the relative variances obtained with the several equations; an arbitrary value of 1 was assigned to the variance of the equation showing the least deviation of any of the equations for a specific system. Thus the relative variance for each equation becomes $\sum R_i^2 / \min \sum R_i^2$. The number of data points was usually such that a ratio of variances less than 2 indicated no statistically significant difference in the ability of the equations to fit a given set of data at a 90% confidence level. Systems that exhibited negative or slightly positive deviations from Raoult's law are listed in part a of Table 5.8. For such systems there is no basis for choice among the two-constant activity-coefficient equations. All the equations do a fairly good job of representing such systems, and the incidence of best fit appears to be fairly random.

Part b of Table 5.8 lists systems that exhibit moderate positive deviations from Raoult's law. For these systems as well it appears to make little difference which of the equations for activity coefficients is used. One can notice, however, that in part b of this table the incidence of best fit is quite high for the Wilson equations as compared with the other three equations.

Part c of Table 5.8 lists systems that exhibit large positive deviations. In almost every case in which the two components are completely miscible the order of preference of the equations is Wilson, Van Laar, Margules, and Scatchard–Hamer. For such systems these differences have very definite statistical significance, leading very strongly to the conclusion that the Wilson equations are the best of the equations tested to use with systems that do not separate into two liquid phases. This difference is especially important in systems with large positive deviations.

A similar analysis was done with the three-constant equations for certain selected difficult-to-fit systems. The results are shown in Table 5.9. When two-constant equations were used, most of these systems were fit best by the Wilson equations. It is significant that in some cases the three-constant equations were not able to produce as good a fit as the two-constant Wilson equations had given. This is particularly evident in the systems involving ethanol and hydrocarbons.

Although the results quoted in the foregoing paragraph do not indicate any universal rule regarding the applicability of the various activity-coefficient equations, they do indicate that the most useful of the equations will undoubtedly be the Wilson and the Van Laar equations. In the three-constant form

Table 5.8 Results of Regression Analyses Based on ln α for Two-Constant Equations

| System | T (°C) | $\sum R_i^2/\min \sum R_i^2$ According to[a] | | | Ref. |
		Wilson (4-68)	Van Laar (4-64)	Margules (4-57)	Scatchard-Hamer (4-59)	
a. Systems with Negative or Slight Positive Deviations (max $\gamma^\infty \leq 1.25$)						
Acetone-chloroform	35.17	1.038	1.215	1.007	1	24
Acetone-methyl acetate	51	1.118	1.118	1.053	1	25
Benzene-chlorobenzene	70	1.180	1.148	1.007	1	26
Butene-1-n-butane[b]	37.8	1.910	4.08	1	—	27
(using Redlich-Kwong	71.1	1.012	1	1.022	—	27
correction)	104.4	2.385	1	2.568	—	27
	137.8	1.068	2.186	1	—	27
n-Butyl chloride-n-butyl bromide	50	1	1.060	1.009	1.006	28
Carbon tetrachloride-cyclohexanol	70	1.001	1	1.020	1.005	29
Carbon tetrachloride-n-heptane	50	1.361	1	1.306	1.120	28
Ethyl bromide-ethyl iodide	30	1.489	1.808	1.048	1	28
b. Systems with Moderate Positive Deviations ($1.25 \leq$ Max $\gamma \leq 10$)						
Acetone-methanol	51	1	1.003	1.016	1.007	25
	100[b]	1	1.015	1.012	—	30
	150[b]	1.013	1	1.575	—	30
Acetone-water	25	1	1.037	1.038	1.232	31
Benzene-aniline	70	1	1.008	1.034	1.033	26
Benzene-anisole	70	1	1.011	1.028	1.020	26
Benzene-benzyl alcohol	70	1	1.515	1.517	1.526	32
Benzene-dioxane	25	1.056	1	1.075	1.079	33
Benzene-n-heptane	60	1.004	1.004	1.007	1	34
	80	1.008	1.007	1	1.007	34
Benzene-i-octane	25	1.043	1.036	1	1.024	35
Benzene-phenol	70	1	1.103	1.332	1.374	32
n-Butyl bromide-n-heptane	50	1.123	1.107	1.105	1	28
n-Butyl chloride-n-heptane	50	1.003	1	1.104	1.045	28
Carbon disulfide-benzene	20	1.045	1.040	1	1.0	36
	25	1	1.641	1.070	1.009	36
	30	1.050	1.045	1	1.043	36
Carbon disulfide-i-butyl chloride	20	1.191	1.190	1.185	1	37
Carbon disulfide-i-pentane	20	1	1.118	1.585	1.123	37
Carbon tetrachloride-n-butanol	50	1.120	1	1.071	1.100	38

Table 5.8 (continued)

System	T (°C)	$\sum R_i^2/\min \sum R_i^2$ According to[a]				Ref.
		Wilson (4-68)	Van Laar (4-64)	Margules (4-57)	Scatchard-Hamer (4-59)	
Chlorobenzene-nitrobenzene	75	1	1.016	1.022	1.072	39
Chlorobenzene-nitropropane	120	1.316	1.285	1.047	1	39
Ethanol-water	25	1	1.126	1.081	1.150	19
	39.76	1.853	1.148	1.065	1	20
	54.81	4.857	1.633	2.473	1	20
	74.79	2.527	1.448	2.228	1	20
Ethyl bromide-ethanol	30	1	1.257	2.090	3.705	38
Ethyl bromide-n-heptane	30	1.019	1	1.448	1.083	28
Ethyl ether-carbon disulfide	20.1	1.090	1.064	1	1.016	40
Ethyl ether-ethanol	25	1.175	1.104	1.104	1	41
Ethyl iodide-ethanol	30	1.012	1	1.560	4.090	38
Hydrogen chloride-ethane	−60	1	1.219	—	—	42
	−50	1	1.119	—	—	42
	−40	1	1.181	—	—	42
	−30	1	1.307	—	—	42
Methyl acetate-methanol	51	1	1.004	1.007	1.368	25
Methanol-water[b]	100	1	1.010	—	—	43
	150	1	1.007	—	—	43
	200	1	1.028	—	—	43
	250	1.005	1	—	—	43
c. Systems with Large Positive Deviations (max $\gamma^\infty > 10$)						
Acrylonitrile-water[c]	25	2.345	1	1.001	1.004	44
	40	3.839	1.016	1	1.013	44
Benzene-benzonitrile	70	1.250	1	2.257	2.288	26
Carbon tetrachloride-ethanol	25	1	3.604	8.699	—	45
	45	1	7.589	16.35	58.91	46
	65	1	25.82	44.34	185.6	46
Ethanol-n-heptane	30	1	21.89	28.57	85.07	47
Ethanol-i-octane	25	1	25.87	25.90	74.63	48
	50	1	22.60	25.55	108.8	48
Ethanol-toluene	35	1	28.70	45.42	169.4	49
	55	1	8.938	11.61	45.82	50
n-Hexane-methanol	45	1	29.53	—	—	51
Methanol-benzene	40	1	3.385	—	—	52

[a] Numbers in parentheses refer to text equation number (Chapter 4).
[b] Vapor-phase corrections to pure components made with Redlich-Kwong equation.
[c] Incomplete solubility in liquid phase, regression analysis based on water-rich solutions.

Table 5.9 Results of Regression Analyses Based on ln α for Three-Constant Equations

		$\sum R_i^2 / \min \sum R_i^2$ According to[a]				
System	T (°C)	Van Laar (Wohl Form) (4–75)	Black (4–80)	Fariss (4–83)	Best 2 Const.	Ref.
Acetone–n-butanol	25	4.71	3.36	1.0	3.16	53
Acetonitrile–water	20	1.17	1.0	1.10	1.01	54
	30	1.0	1.11	1.09	1.11	54
Acrylonitrile–water[b]	25	1.0	1.0	1.0	1.01	44
	40	1.0	1.0	1.01	1.01	44
Benzene–benzyl alcohol	70	1.0	2.05	1.57	1.74	32
Carbon tetrachloride–n-butanol	50	1.01	1.0	1.01	1.01	38
Carbon tetrachloride–ethanol	25	1.0	3.18	2.53	1.26	45
	45	1.0	4.3	3.2	1.1	46
	65	1.06	5.91	4.33	1.0	46
Ethanol–n-heptane	30	3.72	10.5	6.61	1.0	47
Ethanol–i-octane	25	1.39	1.85	1.0	1.03	48
Ethanol–toluene	35	1.17	6.02	3.46	1.0	49
	55	1.0	1.71	1.34	1.17	49, 50
Ethyl bromide–ethanol	30	1.41	1.0	1.07	2.30	38
Ethyl ether–ethanol	25	1.01	1.0	1.0	1.31	41
Ethyl iodide–ethanol	30	1.0	1.25	1.17	5.13	38

[a] Numbers in parentheses refer to text equation number (Chapter 4).
[b] Limited miscibility.

both the Fariss and Wohl modifications of the Van Laar equation appear to be superior to Black's.

Data Analysis Based on Minimal Error in Calculated Pressure. It is often desirable, and sometimes necessary, to obtain the coefficients for the activity-coefficient equations from data consisting only of total pressure, liquid composition, and temperature. It may be difficult in a specific case to obtain a vapor sample that truly represents the composition at equilibrium with the liquid phase. On the other hand, accurate measurements of pressure, temperature, and liquid composition are relatively easy to obtain. These data (π–T–x) can be quite accurate even though the experimental data are obtained under conditions of liquid entrainment, partial or premature vapor condensation, and sampling difficulties. When the volatilities of the components are

5.4 Processing Vapor–Liquid Equilibrium Data

quite different, extremely low concentration of one of the components in the vapor phase adds analytical problems. Small errors in vapor composition under such circumstances can give rise to extremely large errors in the relative volatility and consequently in stage requirements for a distillation column. In any case data on total pressure versus liquid composition at constant temperature are quite adequate for describing completely the vapor–liquid equilibria in a given system. The method of data analysis to exploit such data is described in the following paragraphs.

Equations 3-15 form the basis for the data analysis. These equations can be slightly rearranged to give

$$f_1^\pi y_1 = \frac{(\gamma_1)_L}{(\gamma_1)_G} f_1^* x_1 \exp\left[\frac{(v_1)_L(\pi - p_1^*)}{RT}\right],$$

$$f_2^\pi y_2 = \frac{(\gamma_2)_L}{(\gamma_2)_G} f_2^* x_2 \exp\left[\frac{(v_2)_L(\pi - p_2^*)}{RT}\right]. \tag{5-47}$$

In order to use Equation 5-47 it is advantageous to define a quantity η, which is the ratio of fugacity to pressure for a pure component. By application of Equation 2-31, the definition of η becomes

$$\eta_i \equiv \frac{f_i}{P} = \exp\left\{\frac{1}{RT} \int_0^P \left[(v_i)_G - \frac{RT}{P}\right] dP\right\}. \tag{5-48}$$

The application of the above definition to Equation 5-47 results in

$$\pi y_1 \eta_1^\pi = \frac{(\gamma_1)_L}{(\gamma_1)_G} p_1^* \eta_1^* x_1 \exp\left[\frac{(v_1)_L(\pi - p_1^*)}{RT}\right],$$

$$\pi y_2 \eta_2^\pi = \frac{(\gamma_2)_L}{(\gamma_2)_G} p_2^* \eta_2^* x_2 \exp\left[\frac{(v_2)_L(\pi - p_2^*)}{RT}\right]. \tag{5-49}$$

These two equations can be rearranged and combined to give an expression for the total pressure:

$$\pi = \pi(y_1 + y_2) = \frac{(\gamma_1)_L \eta_1^*}{(\gamma_1)_G \eta_1^\pi} p_1^* x_1 \exp\left[\frac{(v_1)_L(\pi - p_1^*)}{RT}\right].$$

$$+ \frac{(\gamma_2)_L \eta_2^*}{(\gamma_1)_G \eta_2^\pi} p_2^* x_2 \exp\left[\frac{(v_2)_L(\pi - p_2^*)}{RT}\right]. \tag{5-50}$$

There are, of course, numerous special cases of Equation 5-50. If the difference between total pressure and component vapor pressures is small compared with the product RT, the following equation results:

$$\pi = \frac{(\gamma_1)_L \eta_1^*}{(\gamma_1)_G \eta_1^\pi} p_1^* x_1 + \frac{(\gamma_2)_L \eta_2^*}{(\gamma_2)_G \eta_2^\pi} p_2^* x_2. \tag{5-51}$$

If, in addition, the vapor phase forms an ideal solution, the following equation applies:

$$\pi = \left(\frac{\eta_1^*}{\eta_1^\pi}\right)\gamma_1 p_1^* x_1 + \left(\frac{\eta_2^*}{\eta_2^\pi}\right)\gamma_2 p_2^* x_2. \tag{5-52}$$

In the above equation the usual convention has been adopted of omitting the L subscript on the liquid-phase activity coefficient when the vapor phase is assumed to be ideal. If the total pressure and vapor pressures are sufficiently close together, the ratio of η^* to η^π is essentially unity; hence Equation 5-52 can be considerably simplified, to give

$$\pi = \gamma_1 p_1^* x_1 + \gamma_2 p_2^* x_2. \tag{5-53}$$

It should be noted that application of Equation 5-53 does not necessarily require that the vapor phase be an ideal gas, although it is applicable when the vapor phase is an ideal gas.

The parameters for the activity-coefficient equation are determined by a nonlinear regression analysis to minimize the sum of the squares of either (experimental minus calculated pressure), or (logarithm of experimental pressure minus logarithm of calculated pressure), with the pressure calculated according to the appropriate equation above. If one of the Equations 5-50, 5-51, or 5-52 is used to calculate pressure, the pressure calculation must be done by successive numerical approximations since η_1^π and η_2^π are functions of total pressure. However, computation by direct iteration usually converges quite rapidly. In the direct iteration procedure a pressure is assumed from which the values of the η terms are calculated. This value is then substituted into the pressure equation and a value of pressure is calculated. If the calculated value of pressure agrees with the assumed value, the computation is complete; if the two values do not agree, the calculated value of π is used to compute η for the next iteration. Table 5.10 shows the results of an application of the above procedure to the data of Griswold, West, and McMillin [55] for the 1,3-butadiene–furfural system at 100°F. In this application the Wilson equations were used, the vapor phase was assumed to be an ideal solution, and the Redlich–Kwong equation was used to calculate the values of the η terms. Figure 5.25 shows a comparison between experimental pressure and pressure values calculated according to the Wilson equations, using the parameters obtained by regression analysis. Figure 5.26 shows a portion of the x-y diagram calculated by this equation. It is to be noted that the mole fraction furfural in the vapor phase was so small that it would have been quite difficult to obtain y values by analysis of vapor-phase samples.

Analysis of Multistage Data. The basic principles of the use of multistage equipment for measuring vapor–liquid equilibrium are outlined in Section

5.4 Processing Vapor–Liquid Equilibrium Data

Table 5.10 Vapor–Liquid Equilibria for 1,3-Butadiene–Furfural at 100°F Computed from Isothermal π-x Data

Liquid Mole Fraction Butadiene	Pressure Read from Graph in Literature Source (mm Hg)	Pressure Calculated by the Wilson Equations[a]	Vapor Mole Fraction Furfural by the Wilson Equations[a]
0.046	517	537.6	0.0109
0.071	776	782.5	0.0075
0.097	1034	1009.3	0.0058
0.13	1293	1261.5	0.0046
0.17	1551	1521.8	0.0038
0.22	1809	1789.9	0.0032
0.29	2069	2082.6	0.0027
0.38	2328	2356.9	0.0023
0.51	2586	2625.3	0.0019

[a] Equations 4-68. The parameters for the Wilson equations: $A_{12} = 0.53685$; $A_{21} = 0.21779$.

5.3.2. Equation 5-18 applies for calibration runs, whereas Equation 5-17 applies for experimental determination of relative volatility. The application of these equations is demonstrated in the following three example problems.

Example 5-3

A 2-inch-diameter laboratory sieve-tray distillation column is to be calibrated at a nominal pressure of 150 psig. The system n-butane–isobutane has been chosen for the calibration runs. The relative volatility of isobutane with respect to n-butane is 1.23 at this pressure. During the calibration runs the boilup rate was determined by observing the flow rate of liquid leaving the bottom tray of the column. The column was operated at total reflux. The data shown in Table 5.11 were obtained. Determine the number of theoretical plates in the column.

Solution. According to Equation 5-18,

$$N = \frac{\log\{[x_0(1 - x_B)]/[x_b(1 - x_0)]\}}{\log \alpha}.$$

For the first data point this becomes

$$N = \frac{\log[(0.675 \times 0.669)/(0.331 \times 0.225)]}{\log(1.23)} = 6.95.$$

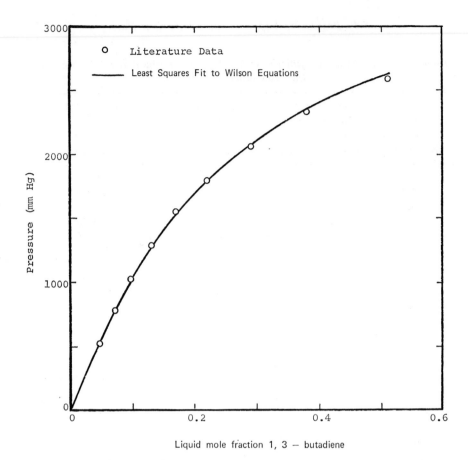

Figure 5.25 Pressure versus liquid composition for 1,3-butadiene–furfural at 100°F

Table 5.11 Data from Calibration of Laboratory Distillation Column

Boilup Rate (ml liquid sec)	Mole Fraction Overhead (x_o)	Isobutane in Bottoms (x_b)	Temperature (°C)	
			Overhead	Reboiler
4.80	0.675	0.331	170	180
5.95	0.645	0.305	171	180
3.75	0.641	0.324	171	180
1.37	0.634	0.327	171	180
3.57	0.607	0.330	171	180
4.14	0.643	0.324	170	180
2.90	0.611	0.329	171	180
2.02	0.634	0.326	171	180

5.4 Processing Vapor–Liquid Equilibrium Data

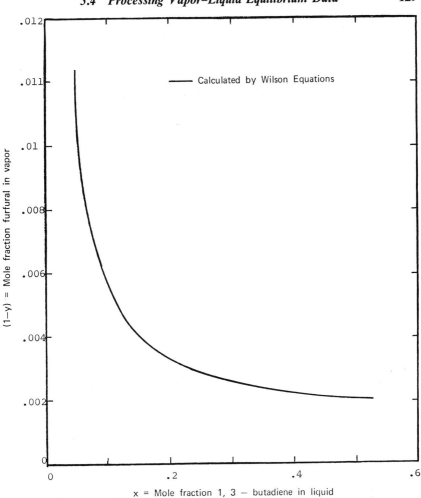

Figure 5.26 Vapor-composition curves for 1,3-butadiene–furfural at 100°F.

A similar calculation was done on each of the data points. These results, arranged in order of increasing boilup rate, are shown in Table 5.12. Since the results indicate more scatter than trend in the data, an average value of 6.12 theoretical trays is considered to characterize the column.

Example 5-4

The column calibrated in Example 5-3 was operated on the system propylene–propane at an overhead pressure of 164.9 psia. From the data

Table 5.12 Number of Theoretical Stages in Laboratory Column

Boilup Rate (ml liquid/sec)	Number of Theoretical Stages
1.37	6.15
2.02	6.16
2.90	5.61
3.57	5.51
3.75	5.80
4.14	5.84
4.80	6.95
5.95	6.92
	Total 48.94
	Average 6.12

shown in Table 5.13 calculate the relative volatility of propylene with respect to propane.

Solution. The relative volatility is calculated by using Equation 5-17, which is illustrated for the first data point:

$$\alpha = \left(\frac{x_o}{1-x_o} \frac{1-x_b}{x_b} \right)^{1/N} = \left(\frac{0.414}{0.586} \frac{0.775}{0.225} \right)^{1/6.12} = 1.156$$

Table 5.13 Propylene–Propane Separation in Laboratory Column

Boilup Rate (ml liquid/sec)	Mole Fraction C_3H_6 in		Temperature (°C)	
	Overhead (x_o)	Bottoms (x_b)	Overhead	Bottoms
4.695	0.414	0.225	83	86
3.204	0.423	0.227	83	86
1.645	0.437	0.225	83	86
1.098	0.460	0.221	82	86

5.4 Processing Vapor–Liquid Equilibrium Data

The results for all the data are tabulated below.

Boilup Rate (ml liquid/sec)	α
4.695	1.156
3.204	1.159
1.645	1.173
1.098	1.196
Total	4.684
Average	1.171

Example 5-5

A 1-inch-diameter Oldershaw column has been calibrated at atmospheric pressure with the benzene–toluene system and found to have an efficiency of 65%. A benzene product stream from the plant contains an impurity known only by its appearance as a peak on a chromatograph. In a total-reflux Oldershaw distillation, using a 10-plate column, the area of the impurity peak on the chromatograph relative to the benzene area was one unit in the overhead and two units in the bottoms. Product specifications can only allow an area of 0.1 for the impurity. From this information determine the relative volatility of the impurity with respect to benzene and discuss the problem of removing the impurity by distillation. The weight fraction of a material is generally proportional to the area of the chromatograph peak.

Solution. Since there are 10 actual plates in the Oldershaw column and the efficiency is 65%, the number of theoretical trays is 6.5. Equation 5-17 is used to calculate the relative volatility. However, since the identity of the impurity is unknown, its molecular weight and its mole fraction in the overhead and bottoms are unknown. The weight fraction, however, can be expressed by the equation

$$w = kA,$$

where w is the weight fraction, k is a constant, and A is the area of the chromatograph peak.

The mole fraction of the impurity can be expressed as follows:

$$x_I = \frac{Mw}{Mw + M_B(1-w)} = \frac{Mw}{(M - M_B)w + M_B},$$

where M is the molecular weight of the impurity and M_B is the molecular weight of benzene.

Since the weight fraction of the impurity is much less than unity,

$$x_I \simeq \frac{M}{M_B} w = \frac{MkA}{M_B}.$$

Since the mole fraction of benzene is very nearly unity in both overhead and bottoms, Equation 5-17 can be expressed as follows:

$$\alpha = \left\{\frac{[Mk(1)/M_B] \times 1}{1 \times [Mk(2)/M_B]}\right\}^{1/6.5} = (0.5)^{0.154} = 0.897.$$

The minimum number of stages required to reduce the peak from an area of one unit to an area of one-tenth of a unit can be calculated by application of Equation 5-18:

$$N = \frac{\log[(0.1/1) \cdot (1/1)]}{\log(0.897)} = 21.7.$$

At total reflux 21.7 theoretical stages will be required to effect the required separation. At a finite reflux this value would be substantially higher. Consequently this separation will be quite expensive to do by distillation. The number of distillation stages is rather large, and in addition virtually the entire feed stream has to be taken as product overhead because the impurity is less volatile than the product. Thus the heat requirement will be large.

Analysis of Infinite-Dilution Activity-Coefficient Data. The significance of the values of the activity coefficients at infinite dilution is discussed briefly in Section 4.3.4. For the Margules, Van Laar, and Scatchard–Hamer equations a very simple relationship exists between the constants of the equation and the infinite-dilution activity coefficients:

$$\begin{aligned} A_{12} &= \ln \gamma_1^\infty, \\ A_{21} &= \ln \gamma_2^\infty. \end{aligned} \quad (5\text{-}54)$$

When any of these equations is modified in the manner outlined by Wohl, the rather simple relationship of Equation 5-54 is maintained, even though additional constants are introduced into the equation. The preceding statement also holds true for the Fariss modification of the Van Laar equation; it does not hold for the Black method of modification. The relationship between the constants of the Wilson equations and the infinite-dilution activity coefficients is not as simple as that expressed in Equation 5-54, although the values of the infinite-dilution activity coefficients do uniquely define the isothermal constants of the Wilson equations as follows:

$$\begin{aligned} \ln \gamma_1^\infty &= 1 - \ln A_{12} - A_{21}, \\ \ln \gamma_2^\infty &= 1 - \ln A_{21} - A_{12}. \end{aligned} \quad (5\text{-}55)$$

5.4 Processing Vapor–Liquid Equilibrium Data

Equations 5-54 and 5-55 indicate that an experimental measurement of the activity coefficients at infinite dilution is of considerable value in correlating vapor–liquid equilibrium data; in fact, if a two-constant equation is adequate, such measurements completely define the system. The thermodynamic basis for direct experimental measurement of the infinite-dilution activity coefficients was first presented by Gautreaux and Coates [56]. Equation 5-51, with the gas phase assumed to form an ideal solution, can be written as

$$\pi = \frac{\gamma_1 \eta_1^* p_1^* x_1}{\eta_1^\pi} \exp\left[\frac{(v_1)_L(\pi - p_1^*)}{RT}\right] + \frac{\gamma_2 \eta_2^* p_2^* x_2}{\eta_2^\pi} \exp\left[\frac{(v_2)_L(\pi - p_2^*)}{RT}\right]. \quad (5\text{-}56)$$

Equation 5-56 can be differentiated with respect to x_1 at constant temperature to give

$$\frac{d\pi}{dx_1} = p_1^* \eta_1^* \exp\left[\frac{(v_1)_L(\pi - p_1^*)}{RT}\right] \left[\frac{\gamma_1 x_1 (v_1)_L}{\eta_1^\pi RT} \frac{d\pi}{dx_1} - \frac{\gamma_1 x_1}{(\eta_1^\pi)^2} \frac{d\eta_1^\pi}{d\pi} \frac{d\pi}{dx_1}\right.$$
$$\left. + \frac{x_1}{\eta_1^\pi} \frac{d\gamma_1}{dx_1} + \frac{\gamma_1}{\eta_1^\pi}\right] + p_2^* \eta_2^* \exp\left[\frac{(v_2)_L(\pi - p_2^*)}{RT}\right]$$
$$\times \left[\frac{\gamma_2 x_2 (v_2)_L}{\eta_2^\pi RT} \frac{d\pi}{dx_2} - \frac{\gamma_2 x_2}{(\eta_2^\pi)^2} \frac{d\eta_2^\pi}{d\pi} \frac{d\pi}{dx_2} + \frac{x_2}{\eta_2^\pi} \frac{d\gamma_2}{dx_2} + \frac{\gamma_2}{\eta_2^\pi}\right] \frac{dx_2}{dx_1}.$$
$$(5\text{-}57)$$

If we let x_1 approach zero and use the following information

$dx_2/dx_1 = -1$,
$d\gamma_1/dx_1$ is finite at $x_1 = 0$,
$\gamma_2 \to 1$,
$x_2 \to 1$,
$d\gamma_2/dx_2 \to 0$ (by application of Equation 5-27),
$\pi \to p_2^*$,

Equation 5-57 becomes

$$\left(\frac{d\pi}{dx_1}\right)^\infty = \frac{\gamma_1^\infty p_1^* \eta_1^*}{\eta_1^{(p_2^*)}} \exp\left[\frac{(v_1)_L(p_2^* - p_1^*)}{RT}\right]$$
$$+ p_2^* \left\{\left[\frac{(v_2)_L}{RT} - \frac{1}{\eta_2^*} \frac{d\eta_2^\pi}{d\pi}\right] \left(\frac{d\eta_2^\pi}{d\pi}\right)^\infty - 1\right\}. \quad (5\text{-}58)$$

Equation 5-58 can now be solved for γ_1^∞:

$$\gamma_1^\infty = \frac{\frac{\eta_1^{(p_2^*)}}{\eta_1^*} \left\{p_2^* \left[1 - \frac{(v_2)_L}{RT} + \frac{1}{\eta_2^*} \frac{d\eta_2^\pi}{d\pi}\left(\frac{d\pi}{dx_1}\right)^\infty\right] + \left(\frac{d\pi}{dx_1}\right)^\infty\right\}}{p_1^* \exp\left[\frac{(v_1)_L(p_2^* - p_1^*)}{RT}\right]}. \quad (5\text{-}59)$$

Equation 5-59 includes the assumption that the gas phase is an ideal solution. It can be further simplified if the gas phase is an ideal gas, in which case the η terms are all unity and the rate of change of η with respect to pressure is zero. The use of this ideal-gas assumption gives

$$\gamma_1^\infty = \frac{p_2^* + \left[1 + \dfrac{p_2^*(v_2)_L}{RT}\right]\left(\dfrac{d\pi}{dx_1}\right)^\infty}{p_1^*} \exp\left[\frac{(v_1)_L(p_1^* - p_2^*)}{RT}\right]. \qquad (5\text{-}60)$$

If, in addition, liquid volumes are small, a further simplification can be made:

$$\gamma_1^\infty = \frac{p_2^* + (d\pi/dx_1)^\infty}{p_1^*}. \qquad (5\text{-}61)$$

Equations 5-59, 5-60, and 5-61 all apply to isothermal data. They can be used with isobaric data if one recognizes that for a binary vapor–liquid system at equilibrium the three variables temperature, pressure, and mole fraction of component 1 are always interdependent; that is, regardless of the type of process considered, a relationship of the following form exists:

$$f(T, \pi, x_1) = 0.$$

Whenever such a functional relationship exists between three variables, the following relationship applies to the partial derivatives:

$$\left(\frac{\partial \pi}{\partial x_1}\right)_T = -\left(\frac{\partial \pi}{\partial T}\right)_{x_1}\left(\frac{\partial T}{\partial x_1}\right)_\pi.$$

Evaluated at $x_1 = 0$, this relationship becomes

$$\left(\frac{d\pi}{dx_1}\right)_T^\infty = -\frac{dp_2^*}{dT}\left(\frac{dT}{dx_1}\right)_P^\infty, \qquad (5\text{-}62)$$

where the subscripts T and P refer to isothermal data and isobaric data, respectively.

Thus Equations 5-59, 5-60, and 5-61 can be used to obtain the value of γ_1^∞ from either isobaric or isothermal data. If isothermal data are used, one needs only to know the properties of the pure components and to determine the slope of the curve for pressure versus liquid composition at $x_1 = 0$. If isobaric data are used, the slope of the curve for temperature versus liquid composition must be determined at $x_1 = 0$. This value is then used in Equation 5-62 to determine the corresponding slope of the curve for pressure versus liquid composition for the equivalent isothermal data.

In theory the data taken in any type of apparatus can be used for determining the infinite-dilution activity coefficients. Actually most sets of data do not give sufficient definition to the curves for pressure versus liquid

5.4 Processing Vapor-Liquid Equilibrium Data

composition or temperature versus liquid composition in the region of infinite dilution to allow the activity coefficients to be determined in the manner just described. An example of such data follows.

Example 5-6

Determine, from the data for pressure versus liquid composition in Table 5.1 the values for the two infinite-dilution activity coefficients of the ethanol–water system at 39.76°C. Compare these values with the values obtained by fitting the relative volatilities to the Van Laar two-constant equation (2.732 and 7.11 for infinite-dilution activity coefficients of water and ethanol, respectively).

Solution. Figure 5.27 shows plots of the pressure versus liquid composition in the dilute-solution regions. Plot *a* represents the ethanol-rich end and plot *b* shows the water-rich end of the composition range. Plot *a* shows the pressure going through a maximum near the lowest composition reported; this makes it very difficult to estimate the slope at infinite dilution, but the slope of 44 mm Hg was obtained from a reasonable curve drawn by eye through the data points. Since this system fits the assumptions involved in Equation 5-61, the activity coefficient can be computed from this slope as follows:

$$\gamma_2^\infty = \frac{129.8 + 44}{54.3} = \frac{173.8}{54.3} = 3.261. \quad vs. \ 2.732$$

The limiting slope to the curve drawn by eye in plot *b* has a value of 445 mm Hg. The activity coefficient determined from this slope is as follows:

$$\gamma_1^\infty = \frac{54.3 + 445}{129.8} = \frac{499.3}{129.8} = 3.85. \quad vs. \ 7.11$$

These values disagree considerably with those obtained by fitting the Van Laar equation to all the data. This discrepancy is undoubtedly due to the inherent difficulty of estimating the slopes when the experimental data are taken at compositions too far removed from infinite dilution. The dotted line in plot *b* indicates the slope that would have been required to agree with the value indicated by the Van Laar equation fit to the data. It is fairly evident that it would have been possible to draw just as reasonable a curve through the data having this slope as it was to obtain the slope indicated by the solid line. The conclusion to be reached from this example is that in order to measure activity coefficients at infinite dilution the data must be taken at low enough mole fractions to allow one to make an accurate estimate of the limiting slope of the curve for pressure versus liquid composition.

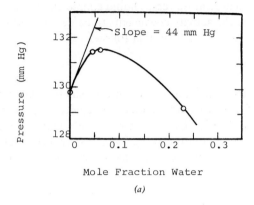

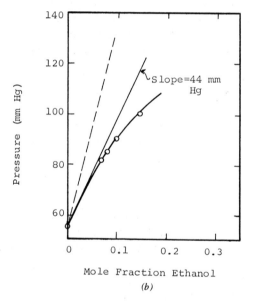

Figure 5.27 Determination of γ^∞ for the ethanol–water system at 39.76°C. (*a*) Ethanol-rich end of the composition range; (*b*) water-rich end.

5.4 Processing Vapor–Liquid Equilibrium Data

When enough data have been taken in the dilute-solution region, it is possible to obtain accurate values of infinite-dilution activity coefficients as indicated in the following example:

Example 5-7

The data shown in Table 5.14 were taken for the acrylonitrile–water system. Determine from these data the infinite-dilution activity coefficient of the acrylonitrile in water.

Table 5.14 Data on Pressure versus Liquid Composition for Acrylonitrile–Water System at 25°C [44]

Mole Percent Acrylonitrile in Liquid	Pressure (mm Hg)
0	23.8
0.12	29.6
0.26	34.2
0.35	39.4
0.51	44.5
0.64	48.2
0.89	55.3
1.08	67.4
1.32	74.3
1.54	81.4
1.72	86.5
1.98	94.2
2.01	97.6
2.07	96.9
2.14	106.9
2.33	103.6
2.44	103.8
2.51	108.0
2.58	109.4

Solution. Figure 5.28 shows the solution to this example. A few of the most dilute data points are shown plotted in this figure, from which a limiting slope of 5040 mm Hg was obtained. This value, used with Equation 5-61, gives an infinite-dilution activity coefficient of 44.3. All of the data shown in Table 5.14 were fitted to a Wohl-type three-constant Van Laar equation,

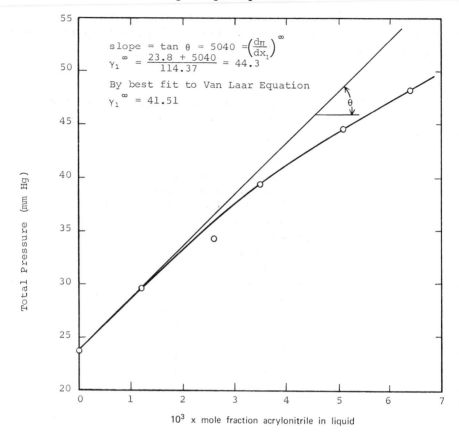

Figure 5.28 Determination of γ^∞ for acrylonitrile–water at 25°C.

which indicated a value of 41.51, giving the least-squares fit to the data. The accuracy of the estimate obtained by the limiting-slope method for this system is very much in contrast to the accuracy obtained in the preceding example. This difference occurs simply because the data in this example are suitable for the use of the limiting-slopes method, whereas the data in the preceding example were not.

The following example shows the application of Equation 5-61 to isobaric data:

Example 5-8

Determine, from the data given in Table 5.4, the values of the infinite-dilution activity coefficients for the isopropanol–water system at the four pressures indicated.

5.4 Processing Vapor–Liquid Equilibrium Data

Solution. Figure 5.29 shows the necessary plots for the atmospheric data. Similar plots were obtained from each set of data. Table 5.15 lists the values of the limiting slopes of the curves for temperature versus liquid composition that were obtained.

Equation 5-62 must be used to convert these limiting temperature slopes to limiting pressure slopes. Use of Equation 5-62 requires a knowledge of the

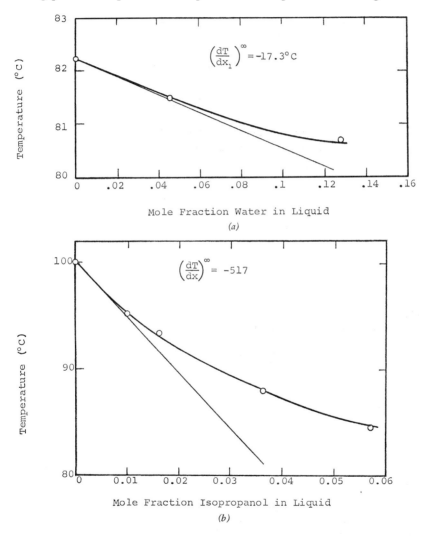

Figure 5.29 Determination of γ^∞ from isobaric data for the isopropanol–water system at 760 mm Hg.

Table 5.15 Limiting Slopes of the Curves for Temperature versus Liquid Composition in Figure 5.29

Pressure	Water-Dilute End		Isopropanol-Dilute End	
(mm Hg)	$T(°C)$	$\left(\dfrac{dT}{dx}\right)^\infty$	$T(°C)$	$\left(\dfrac{dT}{dx}\right)^\infty$
95	38.05	−7.5	50.71	−261.5
190	51.36	−11.8	65.29	−337.5
380	66.02	−14.69	81.68	−403.25
760	82.25	−17.3	100	−517

slopes of the vapor-pressure curves of the two pure components. The Antoine equations for vapor pressure of these two pure components are as follows:

Water:

$$\log p^* = 8.10765 - \frac{1750.286}{t + 235}. \qquad = A - \frac{B}{t+C}$$

Isopropanol:

$d\ln p = \frac{1}{p}\frac{dp}{dt}$

$$\log p^* = 8.39424 - \frac{1730}{t + 231.45}.$$

The slope of the curve for vapor pressure versus temperature in terms of the constants of the Antoine equation and the vapor pressure itself is

$$\frac{dp^*}{dT} = \frac{dp^*}{dt} = \frac{2.303\, p^* B}{(t + C)^2}.$$

The above equation is applied to the atmospheric data as follows:

Water:

$$\frac{dp^*}{dt} = \frac{2.303 \times 760 \times 1750.286}{(100 + 235)^2} = 27.28 \text{ mmHg}/°C.$$

Isopropanol:

$$\frac{dp^*}{dt} = \frac{2.303 \times 760 \times 1730}{(82.25 + 231.45)^2} = 30.87 \text{ mmHg}/°C.$$

The application of Equation 5-62 to the atmospheric data gives the following results:

5.4 Processing Vapor–Liquid Equilibrium Data

Water dilute:

$$\left(\frac{d\pi}{dx_1}\right)^\infty = -30.87 \times (-17.3) = 533.5 \text{ mm Hg.}$$

Isopropanol dilute:

$$\left(\frac{d\pi}{dx_1}\right)^\infty = -27.28 \times (-517) = 14{,}110 \text{ mm Hg.}$$

Equation 5-61 can now be applied:

Water:

$$\gamma_1^\infty = \frac{760 + 533.5}{p_1^*},$$

$$p_1^* = 389.1 \text{ mm Hg},^\dagger$$

$$\gamma_1^\infty = \frac{760 + 533.5}{389.1} = 3.322.$$

Isopropanol:

$$\gamma_1^\infty = \frac{760 + 14{,}110}{1494} = 9.96.$$

The complete results are tabulated in Table 5.16 and plotted in Figure 5.30.

Table 5.16   Results for Example 5-8

Isopropanol Dilute					Water Dilute				
T (°C)	$\dfrac{dp_2^*}{dT}$	$\left(\dfrac{d\pi}{dx_1}\right)^\infty$	p_1^*	γ_1^∞	T (°C)	$\dfrac{dp_2^*}{dT}$	$\left(\dfrac{d\pi}{dx_1}\right)^\infty$	p_1^*	γ_1^∞
50.71	4.69	1,226	182.9	7.22	38.05	5.21	39.09	49.72	2.698
65.29	8.49	2,865	365.2	8.36	51.36	11.36	139.6	95.45	3.451
31.88	15.31	6,175	738.8	8.87	66.02	17.11	251.2	195.9	3.221
100	27.28	14,110	1494	9.96	82.25	30.87	533.5	389.1	3.322

The preceding examples were all based on data obtained with fairly standard equipment and reported in the literature. Much more accurate data can be obtained in apparatus designed specifically for the purpose of using the

† (Calculated from the Antoine equation for water).

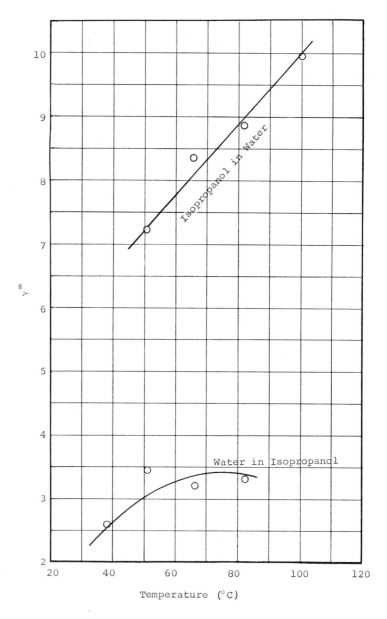

Figure 5.30 Infinite-dilution activity coefficients for isopropanol–water as functions of temperature.

5.4 Processing Vapor-Liquid Equilibrium Data

limiting-slope method for obtaining activity coefficients at infinite dilution. Such an apparatus is based on the modification of the Swietoslawski ebulliometer shown in Figure 5.31. Three or more of these ebulliometers may be arranged as shown in Figure 5.32, all connected to a common pressure-control system. When the apparatus is used in this arrangement, the pure solvent is charged to one of the ebulliometers and very dilute solutions are

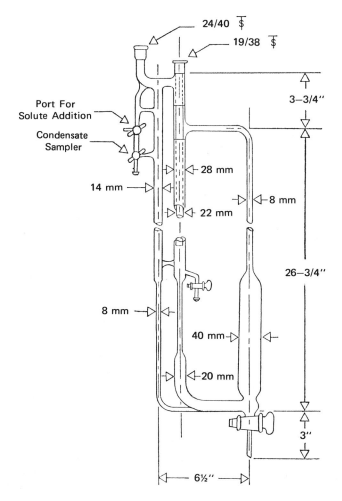

Figure 5.31 Pyrex ebulliometer modified for measurements of infinite-dilution activity coefficients.

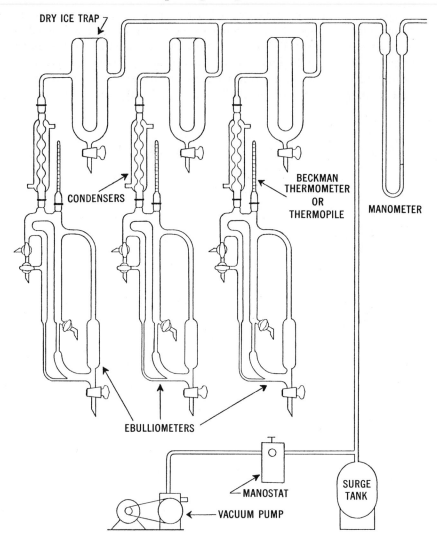

Figure 5.32 Assembly of ebulliometer system.

charged to the remaining ebulliometers, each charge having a different concentration of solvent. Since extremely small temperature differences can be detected between pure solvent and very dilute solution with the temperature-measurement method indicated, very dilute solutions may be used, thus increasing the accuracy of the slope determination. Typical data obtained in such an apparatus are shown in Figure 5.33.

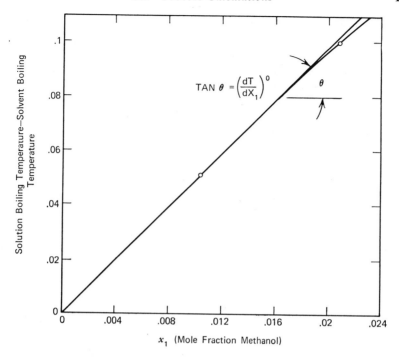

Figure 5.33 Temperature difference versus mole fraction for measurement of infinite-dilution activity coefficients.

5.5 PROCESS CALCULATIONS

Three basic classes of vapor–liquid equilibrium calculations are used extensively in process design. The first is the bubble-point calculation, which consists of the calculation of the equilibrium pressure (or temperature) and vapor composition when temperature (or pressure) and liquid composition are given. The second basic type of calculation is the dew-point calculation. In this case from the given vapor composition the liquid composition and total pressure (or temperature) at a given temperature (or pressure) are to be calculated. The third important basic calculation is the flash calculation. For this type of calculation the overall composition of a mixture is specified, as well as the temperature and pressure. From this information one is required to calculate the fraction of the material that will appear in the liquid phase, and in the vapor phase, and the composition of each of the individual phases.

Although other, more complex, types of vapor–liquid equilibrium calculation may be encountered, they can generally be broken down into a combination of the three basic types indicated above. Thus, by directing our attention to the solution of these three basic types of problem, we can visualize the principles involved in all of the vapor–liquid equilibrium calculations that might be encountered. In all the process calculations it is assumed that the data analysis has been completed and γ values are known as functions of temperature, pressure, and composition.

5.5.1 Bubble-Point Calculations

There are two types of bubble-point calculation. The first (and simpler) type has the following general statement:

Given a multicomponent liquid mixture of composition $x_1, x_2, \ldots, x_n$, where n is the number of components, and given the temperature of the mixture, calculate the pressure at which the mixture will boil. Also calculate the composition of the vapor phase that will be generated by this boiling liquid.

The solution of this general problem requires the application of the multicomponent form of Equation 5-50:

$$\pi = \sum_i \frac{(\gamma_i)_L}{(\gamma_i)_G} \frac{\eta_i^*}{\eta_i^\pi} p_i^* x_i \exp\left[\frac{(v_i)_L(\pi - p_i^*)}{RT}\right]. \tag{5-63}$$

The composition of the vapor phase is calculated from

$$y_i = \frac{(\gamma_i)_L}{(\gamma_i)_G} \frac{\eta_i^*}{\eta_i^\pi} \frac{p_i^*}{\pi} x_i \exp\left[\frac{(v_i)_L(\pi - p_i^*)}{RT}\right]. \tag{5-64}$$

The constants of this problem are T, R, $(v_i)_L$, x_i, p_i^*, and η_i^*. The quantity $(\gamma_i)_L$ is usually independent of pressure, and, since the composition of the liquid phase, x_i, is fixed for the problem, $(\gamma_i)_L$ can also usually be treated as a constant. The quantities η_i^π and $(\gamma_i)_G$ that appear on the right-hand side of Equation 5-63 are generally dependent on pressure and vapor composition; hence Equation 5-63 is not generally an explicit equation in π but requires a numerical method of solution. The solution, however, converges rapidly when a direct-iteration method is used. Once Equation 5-63 is solved, Equation 5-64 gives a direct explicit solution for the vapor composition. A general procedure to be followed in the solution of type 1 bubble-point calculations is as follows:

1. From the given temperature and composition calculate the other constants of the problem (vapor pressures, liquid molal volumes, and fugacity ratios of the pure components at their vapor pressures).

5.5 Process Calculations

2. Assume a value for the total pressure π and vapor composition.
3. Calculate the values of η_i^π and $(\gamma_i)_G$ for each component.
4. Calculate a total pressure π by using Equation 5-63.
5. Calculate the vapor composition (y_i for all components) according to Equation 5-64.
6. If the calculated pressure and vapor composition from step 4 are not the same as the assumed, repeat the calculation from step 3, using the calculated values as the new assumed values.

The above procedure is illustrated with a number of examples.

Example 5-9

Calculate the bubble-point pressure and corresponding vapor composition for a mixture of methyl acetate, acetone, and methanol at 50°C, when the liquid composition consists of 33 mole % methyl acetate, 34 mole % acetone, and 33 mole % methanol. The vapor pressures of the pure components are 585.4, 613.7, and 416.9 mm Hg for the methyl acetate, acetone, and methanol, respectively. Assume for this calculation that both the gas phase and the liquid phase are ideal, and that the difference between total pressure and vapor pressure is negligible.

Solution. With the assumptions used in this problem the partial pressure of each component can be calculated by $\pi y = p^* x$.

Component	πy (mm Hg)	y
Methyl acetate	$0.33 \times 585.4 = 194.1$	$\dfrac{194.1}{540.2} = 0.359$
Acetone	$0.34 \times 613.7 = 208.6$	$\dfrac{208.6}{540.2} = 0.386$
Methanol	$0.33 \times 416.9 = \underline{137.5}$	$\dfrac{137.5}{540.2} = \underline{0.255}$
	$\pi = \sum \pi y = 540.2$	Total 1.000

Example 5-10

Literature data [57] show the values for the infinite-dilution activity coefficient in the methyl acetate–acetone–methanol system to be as follows:

Vapor-Liquid Equilibrium

Solute	Solvent	γ^∞
Methyl acetate	Acetone	1.121
Methyl acetate	Methanol	2.912
Acetone	Methyl acetate	1.160
Acetone	Methanol	2.015
Methanol	Methyl acetate	2.781
Methanol	Acetone	1.676

Assume only that the vapor phase behaves as an ideal gas and recalculate Example 5-9 by using the Wilson equations to express the concentration dependence of the activity coefficient.

Solution. With the assumptions used in this example, the solution is based on the following equation:

$$\pi y_i = \gamma_i p_i^* x_i \exp\left[\left(\frac{\pi - p_i^*}{RT}\right)(v_i)_L\right].$$

Since the calculation is not explicit in the πy product, it must be done numerically, and is best done with the aid of a digital computer. Figure 5.34 shows a FORTRAN program for the solution of this problem.

The FORTRAN variable T is the temperature, the array indicated by ANT contains the constants of the Antoine vapor-pressure equation, the array G consists of the activity coefficient at infinite dilution, and the array V represents the liquid volumes of the three pure components.

The calculations in lines 6 through 37 represent a numerical computation of the Wilson equation parameters from the values of the activity coefficients at infinite dilution. The procedure, symbolically, is as follows:

Let γ_{ij}^∞ represent γ_i^∞ in component j. Then let $B_1 = \frac{+\ln A_{ij}}{-\ln \gamma_{ij}^\infty}$, $B_2 = \frac{+\ln A_{ji}}{-\ln \gamma_{ji}^\infty}$ as initial estimates.

The equations to be solved are

$$\ln \gamma_{ij}^\infty = 1 - \ln A_{ij} - A_{ji},$$
$$\ln \gamma_{ji}^\infty = 1 - \ln A_{ji} - A_{ij}.$$

Letting B_1 represent $\ln A_{ij}$ and B_2 represent $\ln A_{ji}$, these equations are transformed to

$$f_1 = 0 = \ln \gamma_{ij}^\infty + B_1 + e^{B_2} - 1,$$
$$f_2 = 0 = \ln \gamma_{ji}^\infty + B_2 + e^{B_1} - 1.$$

```
  0  $IBFTC E1083Y
  1        DIMENSION G(3,3),A(3,3),ANT(3,3),P(3),X(3),Y(3),GAM(3),V(3),
       1   SUMA(3)
  2        READ(5,1000) T
  3        READ(5,1000) ANT
  4        READ(5,1000) G
  5        READ(5,1000) V
  6        DO 3 I=1,3
  7            A(I,I) = 1.
 10        DO 3 J=1,3
 11        IF(J.LE.I) GO TO 3
 14            B1=-ALOG(G(I,J))
 15            B2=-ALOG(G(J,I))
 16      2  F1 = ALOG(G(I,J)) + B1 + EXP(B2) - 1.
 17            F2 = ALOG(G(J,I)) + B2 + EXP(B1) - 1.
 20            DB1=F1-F2*EXP(B2)
 21            DB2=F2-F1*EXP(B1)
 22        IF(ABS(B1+B2).LT.0.05) GO TO 22
 25            DEN=EXP(B1+B2)-1.
 26            DB1=DB1/DEN
 27            DB2=DB2/DEN
 30     22     B1 = B1 + DB1
 31            B2 = B2 + DB2
 32        IF (ABS(DB1)+ABS(DB2).GT.0.00001) GO TO 2
 35            A(I,J) = EXP(B1)
 36            A(J,I) = EXP(B2)
 37      3 CONTINUE
 42        WRITE (6,1001) A
 43        WRITE (6,1002) G
 44        WRITE (6,1009) V
 45        DO 4 I=1,3
 46      4   P(I)=10.**(ANT(1,I) - ANT(2,I)/(T+ ANT(3,I)))
 50        WRITE (6,1005) P
 51    501 READ (5,1000) X
 52        WRITE (6,1003) X
 53        DO 5 I = 1,3
 54            SUMA(I) = 0.
 55        DO 5 J = 1,3
 56      5   SUMA(I) = SUMA(I) + A(I,J)*X(J)
 61        DO 7 I = 1,3
 62            ALNG = 1. - ALOG(SUMA(I))
 63        DO 6 J = 1,3
 64      6   ALNG = ALNG - A(J,I)*X(J)/SUMA(J)
 66      7   GAM(I) = EXP(ALNG)
 70        WRITE (6,1008) GAM
 71            PI = P(1)*X(1) +P(2)*X(2) +P(3)*X(3)
 72        WRITE (6,1010) PI
 73     11     PIC = 0.
 74        DO 12 I = 1,3
 75     12  PIC = PIC + GAM(I)*P(I)*X(I)*EXP(V(I)*(PI-P(I))/(62370.*
      1    (T+ 273.16)))
 77        WRITE (6,1010) PIC
100        IF (ABS((PIC-PI)/PI).LT.(0.1**5)) GO TO 13
103            PI = PIC
104        GO TO 11
105     13 DO 14 I = 1,3
106     14  Y(I) = GAM(I)*P(I)*X(I)*EXP(V(I)*(PI-P(I))/(62370.*
      1    (T +273.16)))/PI
110        WRITE (6,1011) Y
111        GO TO 501
112   1000 FORMAT (G0.0)
113   1001 FORMAT ( 18H1 WILSON CONSTANTS/9F12.5)
114   1002 FORMAT ( 18H0 GAM(INF. DIL.)  /9F12.5)
115   1003 FORMAT ( 12H0 LIQ. M. F./3F12.5)
116   1005 FORMAT ( 18H0 VAPOR PRESSURES /3F12.5)
117   1008 FORMAT ( 24H0 ACTIVITY COEFFICIENTS /3F12.6)
120   1009 FORMAT ( 18H0 LIQUID VOLUMES   /3F12.6)
121   1010 FORMAT ( 12H0 PRESSURE = F12.4)
122   1011 FORMAT ( 12H0 VAPOR M.F. /3F12.6)
123        END
```

Figure 5.34 FORTRAN program for solution of Example 5-10.

Differentiating, we obtain

$$\frac{\partial f_1}{\partial B_1} = 1; \qquad \frac{\partial f_1}{\partial B_2} = e^{B_2};$$

$$\frac{\partial f_2}{\partial B_1} = e^{B_1}; \qquad \frac{\partial f_2}{\partial B_2} = 1.$$

The Newton–Raphson estimate of the adjustment to the values of B_1 and B_2 is obtained by solving

$$-f_1 = \frac{\partial f_1}{\partial B_1} \Delta B_1 + \frac{\partial f_1}{\partial B_2} \Delta B_2,$$

$$-f_2 = \frac{\partial f_2}{\partial B_1} \Delta B_1 + \frac{\partial f_2}{\partial B_2} \Delta B_2,$$

or

$$-f_1 = \Delta B_1 + e^{B_2} \Delta B_2,$$
$$-f_2 = e^{B_1} \Delta B_1 + \Delta B_2$$

for ΔB_1 and ΔB_2. This calculation is repeated until the calculated adjustments meet the criterion $\Delta B_1 + \Delta B_2 < 0.00001$. Then

$$A_{ij} = e^{B_1}; \qquad A_{ji} = e^{B_2}.$$

Lines 42 and 43 represent the printout of the arrays giving A (the values of the A_{ij} terms just calculated), G (the values of the infinite-dilution activity coefficients), and V (the liquid volume).

Lines 45 through 50 represent the calculation of the vapor pressures of pure components via the Antoine equation and a printout of the values calculated.

Lines 51 and 52 designate the input of the values of the mole fractions of the three components in the liquid phase and a printout (confirmation) of these values.

Lines 53 through 70 describe the calculation of the activity coefficients of the three components by using Equation 4-93.

Line 71 is an initial estimate of the total pressure, based on ideal-solution theory.

Lines 73 through 104 represent the numerical calculation of the actual total pressure at the bubble point according to the following method:

Using the assumed value of π on the right-hand side, calculate π by

$$\pi = \sum_i \gamma_i p_i^* x_i \exp\left(\frac{\pi - p_i^*}{RT}\right).$$

If the new π differs from the assumed π by more than 0.001%, repeat by using the new π as the assumed π.

Lines 105 through 110 represent the calculation and printout of the vapor mole fraction based on the final value of total pressure calculated.

Figure 5.35 shows the computer output sheet for the solution to this problem. It is significant that the final pressure attained was over 100 mm Hg higher than that assumed for the initial ideal-solution estimate and that the composition of the vapor phase as calculated for this problem was significantly different from that calculated in Example 5-9 for an ideal liquid solution. The values of pressure calculated at the end of each iteration, which appear on the output sheet, show that the numerical solution for pressure converged after only four successive trials.

Example 5-11

Calculate the bubble-point pressure and vapor composition for a mixture of butene-1 and *n*-butane at 71.1°C, with liquid compositions of 10 and 20 mole % butene-1. It may be assumed that the vapor phase behaves as an ideal solution, with the pure components following the Redlich–Kwong equation of state. The activity coefficients at infinite dilution have been estimated to be 1.0493 for butene-1 and 1.0303 for *n*-butane. Use the Wilson equations to calculate liquid activity coefficients.

Solution. With the assumptions used in this example, the solution is based on the following equation:

$$\pi y_i = \frac{\gamma_i p_i^* \eta_i^*}{\eta_i^\pi} x_i \exp\left[\left(\frac{\pi - p_i^*}{RT}\right)(v_i)_L\right].$$

The calculation must be done by numerical methods, similar to those used in Example 5-10. However, there is the additional problem of calculating the fugacity ratios of the two pure components at the total pressure of the system and at their vapor pressure. This calculation, of course, is best done with the aid of a digital computer. Figure 5.36 shows the FORTRAN program for the solution of this problem.

In the program lines 2 through 10 read in the input data for the pure component. Lines 2 through 5 read in arrays having the same meaning as the corresponding arrays in the solution of Example 5-10. The arrays read in by lines 6 through 10 are as follows:

TC represents the critical temperatures of the two pure components.

PC represents the critical pressures of the pure components.

R represents the gas constant for the units used for the critical temperatures and pressures.

PCONV represents the conversion factor between the units used for critical pressure and those used for vapor pressure in the Antoine equation.

TABS0 represents the negative of the absolute zero temperature in the scale used.

```
WILSON CONSTANTS
 1.00000    0.71891    0.57939    1.18160    1.00000    0.97513    0.52297    0.50878    1.00000
GAM(INF. DIL.)
 1.00000    1.16000    2.78100    1.12100    1.00000    1.67600    2.91200    2.01500    1.00000
LIQUID VOLUMES
83.770000  76.810000  42.050000
VAPOR PRESSURES
585.42289  613.68894  416.89466
LIQ. M.F.
 0.33000    0.34000    0.33000
ACTIVITY COEFFICIENTS
 1.201301   1.029787   1.418105

PRESSURE =   539.4190

PRESSURE =   641.9891

PRESSURE =   642.2138

PRESSURE =   642.2143

VAPOR M.F.
 0.361458   0.334613   0.303930
```

Figure 5.35 Solution of Example 5-10.

```
0 $IBFTC E1083W
1         DIMENSION G(2),A(2),ANT(3,2),P(2),X(2),Y(2),GAM(2),V(2),TC(2),
        1 PC(2),ARK(2),BRK(2),B(2),VG(2),ETA(2),ETAPI(2)
2         READ (5,1000) T
3         READ (5,1000) ANT
4         READ (5,1000) G
5         READ (5,1000) V
6         READ (5,1000) TC
7         READ (5,1000) PC
10        READ (5,1000) R,PCONV,TABSO
11        DO 2 I = 1,2
12      2 B(I) = 0.
14     21 F1 = ALOG(G(1)) + B(1) + EXP(B(2)) - 1.
15        F2 = ALOG(G(2)) + B(2) + EXP(B(1)) - 1.
16        C=100.
17        IF(ABS(B(1)+B(2)).GT.0.05)
        1 C = EXP(B(1) + B(2)) - 1.
22        DB2 = (F2 - F1*EXP(B(1)))/C
23        DB1 = (F1 - F2*EXP(B(2)))/C
24        IF (ABS(DB1)+ ABS(DB2).LT.0.00001) GO TO 22
27        B(1)=B(1)+DB1
30        B(2)=B(2)+DB2
31        GO TO 21
32     22 T = T + TABSO
33        DO 5 I = 1,2
34        A(I) = EXP(B(I))
35        ARK(I) = 0.4278*R**2* TC(I)**(2.5)/PC(I)
36        BRK(I) = 0.0867*R*TC(I)/PC(I)
37        TP = T-TABSO+ANT(3,I)
40        P(I) = 10.**(ANT(1,I)-ANT(2,I)/(TP))
41        PI= PCONV*P(I)
42        VG(I) = R*T/PI
43      3 F = R*T/(VG(I)-BRK(I)) -ARK(I)/(SQRT(T)*VG(I)*(VG(I)+BRK(I))
        1 ) - PI
44        DFDV = ARK(I)*(2.*VG(I))/(SQRT(T)*(VG(I)*(VG(I)+BRK(I)))**2)
        1 - R*T/(VG(I)-BRK(I))**2
45        IF (F.LT.(0.00001*PI)) GO TO 4
50        VG(I) = VG(I) - F/DFDV
51        GO TO 3
52      4 ETA(I) = PI*VG(I)/(R*T) - 1. - ARK(I)*ALOG(1.+BRK(I)/VG(I))
        1 /(BRK(I)*R*T**(2.5)) +
        2 ALOG(R*T/(PI*(VG(I)-BRK(I))))
53      5 ETA(I)= EXP(ETA(I))
55        TD = T - TABSO
56        WRITE (6,1001)TD
57        WRITE (6,1002)ANT
60        WRITE (6,1003)G
61        WRITE (6,1004)V
62        WRITE (6,1005)TC
63        WRITE (6,1006)PC
64        WRITE (6,1007)R,PCONV,TABSO
65        WRITE (6,1008)P
66        WRITE (6,1009)VG
67        WRITE (6,1010)ETA
70    501 READ(5,1000)X(1)
```

Figure 5.36 FORTRAN program for solution of Example 5-11.

```
 71          WRITE(6,1011) X(1)
 72               X(2) = 1. - X(1)
 73          DO 6 I=1,2
 74               J= 2
 75          IF (I.EQ.2)J=1
100               ALNG = 1. - ALOG(X(I)+A(I)*X(J)) - X(I)/(X(I)+A(I)*X(J))
            1          - A(J)*X(J)/(X(J)+A(J)*X(I))
101       6       GAM(I) = EXP(ALNG)
103          WRITE (6,1012) GAM
104               PRESS = P(1)*X(1) + P(2)*X(2)
105       7       PI = PRESS*PCONV
106          WRITE (6,1013) PRESS
107          DO 10 I=1,2
110               VG(I)= R*T/PI
111       8       F = R*T/(VG(I)-BRK(I))-ARK(I)/(SQRT(T)*VG(I)*(VG(I)+BRK(I)))
            1          - PI
112               DFDV = ARK(I)*(2.*VG(I))/(SQRT(T)*VG(I)*(VG(I)+BRK(I)))**2)
            1          - R*T/(VG(I)-BRK(I))**2
113          IF (F.LT.(0.00001*PI)) GO TO 9
116               VG(I) = VG(I) - F/DFDV
117          GO TO 8
120       9       ETAPI(I) = PI*VG(I)/(R*T) - 1. - ARK(I)*ALOG(1.+BRK(I)/VG(I))
            1          /(BRK(I)*R*T**(2.5)) +
            2          ALOG(R*T/(PI*(VG(I)-BRK(I))))
121      10       ETAPI(I) = EXP(ETAPI(I))
123               PIC = 0.
124          DO 12 I = 1,2
125               F1 = EXP(V(I)*(PI-P(I)*PCONV)/(R*T))
126               F2 = GAM(I)*ETA(I)*P(I)*X(I)/ETAPI(I)
127      12       PIC = PIC + F1*F2
131          WRITE (6,1013) PIC
132          IF (ABS((PIC-PRESS)/PIC).LT.(0.1**4)) GO TO 13
135               PRESS = PIC
136          GO TO 7
137      13  WRITE (6,1009) VG
140          DO 14 I= 1,2
141               F1 = EXP(V(I)*(PI-P(I)*PCONV)/(R*T))
142               F2 = GAM(I)*ETA(I)*P(I)*X(I)/ETAPI(I)
143      14       Y(I) = F1*F2/PRESS
145          WRITE (6,1014) Y
146          GO TO 501
147    1000 FORMAT(G0.0)
150    1001 FORMAT(   18H1 TEMPERATURE   =   F10.3)
151    1002 FORMAT(   24H0 ANTOINE COEFFICIENTS  /3G15.6/3G15.6)
152    1003 FORMAT(   42H0 INFINITE DILUTION ACTIVITY COEFFICIENTS /2G15.6)
153    1004 FORMAT(   18H0 LIQUID VOLUMES   / 2G15.5)
154    1005 FORMAT(   24H0 CRITICAL TEMPERATURES /2G15.6)
155    1006 FORMAT(   24H0 CRITICAL PRESSURES    /2G15.6)
156    1007 FORMAT(   6H0 R = G15.6,    12H PCONV  =    G15.6, 9H T CORR =G15.6)
157    1008 FORMAT(   18H0 VAPOR PRESSURES /2G15.6 )
160    1009 FORMAT(   18H0 GAS VOLUMES    /2G15.6)
161    1010 FORMAT(   36H0 FUGACITY RATIOS AT VAPOR PRESSURE /2G15.6)
162    1011 FORMAT(   30H0 MOLE FRACTION COMPONENT 1  /G15.6)
163    1012 FORMAT(   24H0 ACTIVITY COEFFICIENTS /2G15.6)
164    1013 FORMAT(   12H PRESSURE = G15.6)
165    1014 FORMAT(   24H VAPOR MOLE FRACTIONS   /2G15.6)
166          END
```

Figure 5.36 (*Continued*).

5.5 Process Calculations

Statements in lines 11 through 31 represent the determination of the constants of the Wilson equations from the values of the infinite-dilution activity coefficients. This is done in the same manner as illustrated in Example 5-10.

Statements in lines 34 through 42 represent the establishment of the constants to be used in the Redlich–Kwong equation of state for the pure component and the calculation of the vapor pressure of each component. Symbolically the computations carried out on these lines are as follows:

Line	Symbol	Meaning
34	A(1)	Wilson constant A_{ij}
35	ARK	The a of the Redlich–Kwong equation: $$a = \frac{0.4278 R^2 T_c^{2.5}}{P_c}$$
36	BRK	The b of the Redlich–Kwong equation: $$b = \frac{0.0867 R T_c}{P_c}$$
37–40	P	The vapor pressure p^* by the Antoine equation: $\log p^* = A - [B/(T + C)]$, where $A = \text{ANT}(1,\text{I})$ $B = \text{ANT}(2,\text{I})$ $C = \text{ANT}(3,\text{I})$
41	PI	p^* in the units of P_c

Statements on lines 42 through 51 represent a numerical solution for the pure-component gas volume of each component at its vapor pressure, obtained by using the Redlich–Kwong equation of state:

Line 42: VG is v_G. Initial estimate:

$$v_G = \frac{RT}{p^*}.$$

Line 43: F is $p - p^*$ (p calculated by the Redlich–Kwong equation):

$$p = \frac{RT}{v_G - b} - \frac{a}{\sqrt{T} v_G (v_G + b)}.$$

Line 44 calculates dF/dV, represented by DFDV:

$$\frac{dF}{dv} = \frac{2av_G}{\sqrt{T}[v_G(v_G+b)]^2} - \frac{RT}{(v_G-b)^2}.$$

Line 45 tests to determine whether F is sufficiently near zero.

Line 50 estimates the new value of v_G by the Newton–Raphson method:

$$(v_G)_{\text{new}} = (v_G)_{\text{old}} - \frac{F}{(dF/dv_G)}.$$

Line 51 directs program to repeat calculation from line 43.

Statements on lines 52 and 53 are the calculation of the fugacity ratios by using the vapor pressure of the pure component and the volumes just calculated in the preceding step. The fugacity coefficient is calculated by the following equation:

$$\ln \eta^* = \frac{p^* v_G}{RT} - 1 - \frac{a \ln(1+b/v_G)}{bRT^{2.5}} + \ln\left[\frac{RT}{p^*(v_G-b)}\right].$$

Statements on lines 56 through 67 call for a printout of the input data, plus the calculated data for the pure components.

Line 70 requests that the composition of the mixture whose bubble point is to be calculated be read in.

Line 71 directs the printout of a confirmation of the input mole fraction.

Lines 72 through 103 represent the calculation of the activity coefficients of the two components according to the Wilson equations and a printout of the values calculated.

Lines 104 through 136 represent the iterative calculation of the total pressure:

Line 104 makes the initial estimate of pressure:

$$\pi = p_1^* x_1 + p_2^* x_2.$$

Line 105 changes units of π to correspond to the R value used.

Line 106 writes the current estimated value of π.

Line 107 requests the calculation between lines 110 and 121 to be done for each component.

Line 110 initiates calculation of v_G for the pure component at estimated total pressure π. Initial estimate:

$$v_G = \frac{RT}{\pi}.$$

Lines 111 through 117 calculate v_G at total pressure in same manner as outlined above in discussion of lines 42 through 51.

5.5 Process Calculations

Lines 120 through 121 calculate η^π (same equation as for η^*, with p^* replaced by π).

Lines 123 through 127 calculate new π (PIC) by

$$\pi = \sum_i \pi x_i = \sum \frac{\gamma_i p_i^* \eta_i^*}{\eta_i^\pi} x_i \exp\left[\frac{(v_i)_L(\pi - p_i^*)}{RT}\right].$$

Line 131 prints out the current value of π.
Line 132 tests calculated π to determine if it is sufficiently near the assumed π.
Line 135 sets assumed π = calculated π.
Line 136 directs repetition of calculation from line 105.

The calculation and lines 140 through 145 represent the calculation of the value of the mole fraction of component 1 in the vapor, using the results of the iterative calculation concluded previously.

Line 145 represents the printout of the value of y.

Line 146 directs a return to the point at which a new value of the liquid mole fraction is read. The lines following line 146 are instructions for the format for printing out the results.

Figure 5.37 shows the computer output sheet for the solution to this problem. It should be noted that for this example, as well as Example 5-10, only a very few iterations were required for the convergence of the pressure calculation.

Example 5-12

Repeat the problem in Example 5-11, using the Van Laar equation for activity coefficients, the virial equation of state truncated after the second virial coefficient term for pure-component vapor-phase imperfections, and the virial equation for estimation of vapor-phase activity coefficients using the geometric-mean assumption for interaction as outlined in Chapter 4, Equation 4-21. The values of the infinite-dilution activity coefficients to be used are 1.117 and 1.325 for butene-1 and n-butane, respectively.

Solution. The solution to this problem is based on the following equation:

$$\pi y_i = \frac{(\gamma_i)_L \eta_i^* p_i^* x_i}{(\gamma_i)_G \eta_i^\pi} \exp\left[\frac{(v_i)_L(\pi - p_i^*)}{RT}\right].$$

Figure 5.38 shows a FORTRAN program for the solution of this problem on the digital computer.

The input data to be read, indicated in lines 2 through 10 of the program, have the same meaning as the input data of the preceding problem, Example 5-11.

```
TEMPERATURE   =      71.100

ANTOINE COEFFICIENTS
  6.842900         926.1000         240.0000
  6.830290         945.9000         240.0000

INFINITE DILUTION ACTIVITY COEFFICIENTS
  1.049300         1.030300

LIQUID VOLUMES
  105.700          111.900

CRITICAL TEMPERATURES
  419.6000         425.2000

CRITICAL PRESSURES
  39.70000         37.47000

R =     82.06000      PCONV  =      0.131600E-02  T CORR =    273.1600

VAPOR PRESSURES
  7345.877         6162.949

GAS VOLUMES
  2922.258         3483.162

FUGACITY RATIOS AT VAPOR PRESSURE
  1.025909         1.023267

MOLE FRACTION COMPONENT 1
  0.100000

ACTIVITY COEFFICIENTS
  1.035557         1.000675
PRESSURE  =      6281.241
PRESSURE  =      6310.916
PRESSURE  =      6310.916
PRESSURE  =      6311.183

GAS VOLUMES
  3401.495         3401.495
VAPOR MOLE FRACTIONS
  0.120364         0.879679

MOLE FRACTION COMPONENT 1
  0.200000

ACTIVITY COEFFICIENTS
  1.025114         1.002449
PRESSURE  =      6399.534
PRESSURE  =      6448.189
PRESSURE  =      6448.189
PRESSURE  =      6448.636

GAS VOLUMES
  3329.082         3329.082
VAPOR MOLE FRACTIONS
  0.233272         0.766798
```

Figure 5.37   Solution of Example 5.11.

```
 0    $IBFTC E1083V
 1          DIMENSION G(2),A(2),ANT(3,2),P(2),X(2),Y(2),GAM(2),V(2),TC(2),
            1 PC(2),B(2),PP(2)
 2          READ (5,1000) T
 3          READ (5,1000) ANT
 4          READ (5,1000) G
 5          READ (5,1000) V
 6          READ (5,1000) TC
 7          READ (5,1000) PC
10          READ (5,1000) R,PCONV, ABSO
      C
11          DO 2 I = 1,2
12          TR    = (T + ABSO)/TC(I)
13          B(I)  = (0.125*R*TC(I)/PC(I))
            1     *(1.-3.375*(0.396/TR+1.181/TR**2-0.684/TR**3
            2     +0.384/TR**4))
14          P(I)  = 10.**(ANT(1,I)-ANT(2,I)/(T+ANT(3,I)))
15      2   A(I)  = ALOG(G(I))
      C
17          B12   = (ABS(B(1)+B(2))/(B(1)+B(2))*SQRT(ABS(B(1)*B(2)))
            1      -B(1)-B(2))/(R*(T+ABSO))
20          WRITE (6,1001) T
21          WRITE (6,1002) ANT
22          WRITE (6,1003) G
23          WRITE (6,1004) V
24          WRITE (6,1005) TC
25          WRITE (6,1006) PC
26          WRITE (6,1007) R,PCONV, ABSO
27          WRITE (6,1008) P
30          WRITE (6,1009) B,B12
31      501 READ  (5,1000) X(1)
32          WRITE (6,1011) X(1)
33          X(2)  = 1.-X(1)
34          DO 6 I = 1,2
35          J     = 2
36          IF (I .EQ. 2) J = 1
41          Z     = A(J)*X(J)/(A(I)*X(I)+A(J)*X(J))
42          ALNG  = A(I)*Z**2
43      6   GAM(I) = EXP(ALNG)
45          WRITE (6,1012) GAM
46          PI    = P(1)*X(1) + P(2)*X(2)
47      7 WRITE (6,1013) PI
50          PIC   = 0.
51          DO 8 I = 1,2
52          J     = 2
53          IF (I .EQ. 2) J = 1
56          PP(I) = GAM(I)*P(I)*X(I)*EXP((V(I)-B(I))*PCONV*(PI-P(I))
            1      /(R*(T+ABSO)))/EXP(B12*PCONV*PI*Y(J)**2)
57      8   PIC   = PIC + PP(I)
61          DO 9 I = 1,2
62      9   Y(I)  = PP(I)/PIC
64          IF (ABS((PIC-PI)/PI) .LT. (0.1**4)) GO TO 13
67          PI    = PIC
70          GO TO 7
71     13   WRITE (6,1014) Y
72          GO TO 501
73     1000 FORMAT (G0.0)
74     1001 FORMAT(  18H1 TEMPERATURE  =  F10.3)
75     1002 FORMAT(  24H0 ANTOINE COEFFICIENTS   /3G15.6/3G15.6)
76     1003 FORMAT(  42H0 INFINITE DILUTION ACTIVITY COEFFICIENTS /2G15.6)
77     1004 FORMAT(  18H0 LIQUID VOLUMES   / 2G15.5)
100    1005 FORMAT(  24H0 CRITICAL TEMPERATURES / 2G15.5)
101    1006 FORMAT(  24H0 CRITICAL PRESSURES    / 2G15.6)
102    1007 FORMAT(   6H0 R = G15.6,    12H PCONV  =   G15.6,9H T CORR = G15.6)
103    1008 FORMAT(  18H0 VAPOR PRESSURES /2G15.6)
104    1009 FORMAT(  24H0 VIRIAL COEFFICIENTS   /2G15.6/1X,   48H CONSTANT FO
            1R VAPOR PHASE ACTIVITY COEFFICIENT = G15.6)
105    1011 FORMAT(  24H0LIQ. M.F., COMPONENT 1 /G15.6)
106    1012 FORMAT(  30H0 LIQ. ACTIVITY COEFFICIENTS  /2G15.6)
107    1013 FORMAT(  12H  PRESSURE = G15.6)
110    1014 FORMAT(  24H VAPOR MOLE FRACTIONS   /2G15.6)
111         END
```

Figure 5.38 FORTRAN program for solution of Example 5-12.

Lines 11 through 15 represent the calculation of the properties of the pure components, which are indicated symbolically as follows:

Line 12 calculates reduced temperature (TR):

$$T_R = \frac{T}{T_c}.$$

Line 13 calculates the second virial coefficient by the method of Black [58]:

$$B = \frac{RT_c}{8P_c}\left(\frac{0.396}{T_R} + \frac{1.181}{T_R{}^2} - \frac{0.684}{T_R{}^3} + \frac{0.384}{T_R{}^4}\right).$$

Line 14 converts γ^∞ to Van Laar constants:

$$A_{ij} = \ln(\gamma_i^\infty)_L.$$

Line 15 establishes the constant for the vapor-phase activity coefficient:

$$B12 = \left(\frac{B_{12} - B_{11} - B_{22}}{RT}\right),$$

where $B_{12} = \sqrt{B_{11}B_{22}}$ and $\ln(\gamma_i)_G = (B_{12})\pi(1 - y_i)^2$.

Lines 20 through 30 call for the printout of the input and calculated pure-component data.

Line 31 requests that the composition of the liquid phase for the bubble point to be calculated be read in.

Line 32 directs a printout confirmation of the liquid-phase composition.

Lines 33 through 45 represent the calculation of the liquid-phase activity coefficients with the Van Laar equation. This calculation is indicated symbolically as follows:

Lines 34 through 36: establish values of the subscripts i and j.
Line 41: calculate Z_j (Z):

$$Z_j = \frac{A_{ji}x_j}{(A_{ij}x_i + A_{ji}x_j)}.$$

Lines 42 and 43: calculate $(\gamma_i)_L$ [GAM(I)] by $\ln(\gamma_i)_L = A_{ij}Z_j{}^2$.

Line 45 requests a printout of the calculated values of the activity coefficients.

Line 46 establishes the initial estimate for the calculated pressure.

Lines 47 through 70 direct the iterative calculation of the total pressure. This calculation is similar to those used in Examples 5-10 and 5-11. It proceeds as follows:

5.5 Process Calculations

Lines 50 through 57 calculate π (PIC) by

$$\pi = \sum_i \pi y_i = \sum \frac{(\gamma_i)_L p_i^* x_i}{(\gamma_i)_G} \left(\frac{\eta_i^*}{\eta_i^\pi}\right) \exp\left[\frac{(v_i)_L(\pi - p_i^*)}{RT}\right],$$

where

$$\frac{\eta_i^*}{\eta_i^\pi} = \exp\left[\frac{-B_{ii}(\pi - p_i^*)}{RT}\right]$$

and

$$\ln(\gamma_i)_G = (B12)\pi y_i^2.$$

Line 62 calculates y_i as

$$y_i = \frac{\pi y_i}{\pi}.$$

Line 64 tests calculated versus assumed π for convergence.
Line 67 sets π for the next iteration at the value just calculated.
Line 70 directs repetition of the calculation from line 47.
Line 72 requests that a new value of liquid composition be read and a new calculation started from line 31.

Figure 5.39 shows the printout of the solution to Example 5-12. It should be noted that virtually the same results were obtained in Examples 5-7 and 5-12, in spite of the fact that substantially different values of infinite-dilution activity coefficients were used in the input. This effect occurs because nonideal-solution behavior in the vapor phase was ignored in Example 5-12. Since the pressure is relatively high, some nonideal-solution behavior in the vapor phase might be expected. The equation used for partial pressures contained the liquid-phase activity coefficient in the numerator and the gas-phase activity coefficient in the denominator; hence there is some compensation effect between these two activity coefficients. In obtaining values of the infinite-dilution activity coefficients from the experimental data the value obtained for the liquid-phase activity coefficient will attempt to compensate for vapor-phase deviation if the vapor solution is assumed to be ideal. It should also be noted that the values of the activity coefficient in the liquid phase used when vapor-phase nonideal-solution behavior was ignored were nearer unity (representing ideal solution in the liquid phase) than when the nonideal-solution behavior of the vapor phase was considered.

The second general type of bubble-point calculation can be expressed as follows:

Given a liquid composition $x_1, x_2, \ldots, x_n$ and given the total pressure, calculate the temperature at which the mixture will boil and the composition of the vapor that will be generated by it.

```
TEMPERATURE    =     71.100

ANTOINE COEFFICIENTS
   6.842900        926.1000       240.0000
   6.830290        945.9000       240.0000

INFINITE DILUTION ACTIVITY COEFFICIENTS
   1.170000        1.325000

LIQUID VOLUMES
   105.700         111.900

CRITICAL TEMPERATURES
   419.600         425.200

CRITICAL PRESSURES
   39.70000        37.47000

R =    82.06000     PCONV  =    0.131600E-02  T CORR =    273.1600

VAPOR PRESSURES
   7345.877        6162.949

VIRIAL COEFFICIENTS
   -567.0688       -628.2772
CONSTANT FOR VAPOR PHASE ACTIVITY COEFFICIENT =    0.211843E-01

LIQ. M.F., COMPONENT 1
   0.100000

LIQ. ACTIVITY COEFFICIENTS
   1.149365        1.000959
PRESSURE =    6281.241
PRESSURE =    6391.269
PRESSURE =    6295.147
PRESSURE =    6276.451
PRESSURE =    6272.777
PRESSURE =    6272.056
VAPOR MOLE FRACTIONS
   0.113449        0.886551

LIQ. M.F., COMPONENT 1
   0.200000

LIQ. ACTIVITY COEFFICIENTS
   1.128535        1.004225
PRESSURE =    6399.534
PRESSURE =    6379.268
PRESSURE =    6389.076
PRESSURE =    6390.940
VAPOR MOLE FRACTIONS
   0.226298        0.773702
```

Figure 5.39 Solution of Example 5-12.

For this type of problem the only constant quantities are total pressure π and liquid composition x_i. The values of $(\gamma_i)_L$, $(\gamma_i)_G$, p_i^*, η_i^*, η_i^π, and $(v_i)_L$ are all dependent on the temperature that is to be calculated. The value of $(\gamma_i)_G$ is also dependent on vapor composition. The trial-and-error calculation is an order of magnitude more complex than it was for the type 1 calculation. A method of solution that might be programmed for a digital computer is as follows:

5.5 Process Calculations

1. Assume a value for the temperature T.
2. Calculate the values of the parameters that are dependent on temperature but constant with pressure.
3. Calculate a value of π and vapor composition in the same manner as described for the type 1 calculation.
4. Check the value of π calculated against the value of π given; if the calculated value is less than the given value, increase the value of T, but if it is greater than the given value, decrease the value of T.
5. Repeat from step 2 until a value of π calculated is found to agree with the value of π given.

If a computer is not available, the same procedure may be followed except that the values of π calculated would be plotted against the temperature T. From such a plot one can read the value of temperature at which π calculated is equal to π given. The type 2 calculation is illustrated in the following example:

Example 5-13

Using the same data as in Example 5-12, calculate the temperature at which solutions containing 10 and 20 mole % butene-1 will boil at a pressure of 10 atmospheres. Assume that the infinite-dilution activity coefficients do not change with temperature over the range of the calculation.

Solution. The solution to this problem is obtained by applying the program used in Example 5-12 at a number of different temperatures. The results of calculations of this type are shown in Table 5.17.

Table 5.17 Results of Calculation for Example 5-13

T (°C)	Pressure (mm Hg)		y_1	
	$x_1 = 0.1$	$x_1 = 0.2$	$x_1 = 0.1$	$x_1 = 0.2$
71.1	6272	6391	0.1134	0.2263
72	6398	6518	0.1132	0.2260
75	6831	6953	0.1124	0.2248
78	7284	7408	0.1116	0.2236
80	7597	7722	0.1111	0.2229

Solutions read from Figures 5.40 and 5.41:

$x_1 = 0.1$	$x_1 = 0.2$
$T = 80.02$	$T = 79.3$
$y_1 = 0.1110$	$y_1 = 0.2232$

These results are shown plotted in Figures 5.40 and 5.41. From these plots the solution for pressure is read by finding the temperature value that corresponds to a total pressure of 7600 mm Hg. The vapor mole fraction is read from Figure 5.41, using the temperature obtained from Figure 5.40.

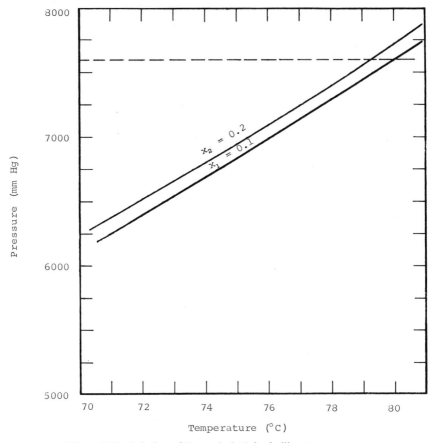

Figure 5.40 Solution of Example 5-13 for boiling temperature

5.5.2 Dew-Point Calculations

Dew-point calculations are in many respects quite similar to bubble-point calculations. There are two types of dew-point calculation, corresponding to the two types of bubble-point calculation. The general statement of the first type of dew-point calculation is as follows:

5.5 Process Calculations

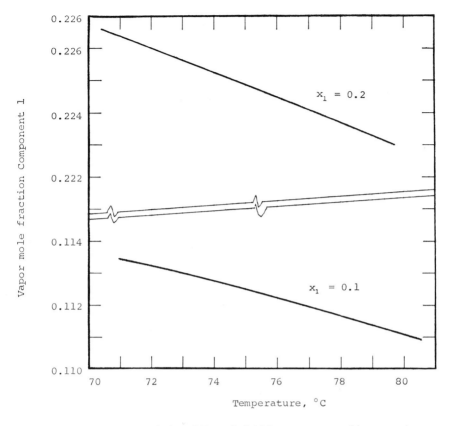

Figure 5.41 Solution of Example 5-13 for vapor composition.

Given a vapor mixture containing n components and having the composition $y_1, y_2, \ldots, y_n$ and given the temperature of the mixture, calculate the pressure at which condensation will begin. Also calculate the composition of the liquid phase that will be condensed from this vapor.

The solution of the general problem requires the application of Equation 5-63 and a rearranged form of Equation 5-64:

$$x_i = \frac{(\gamma_i)_G}{(\gamma_i)_L} \frac{\eta_i^\pi}{\eta_i^*} \frac{\pi}{p_i^*} y_i \exp\left[\frac{(v_i)_L(p_i^* - \pi)}{RT}\right]. \tag{5-65}$$

The constants of this problem are T, R, y_i, p_i^*, η_i^*, and $(v_i)_L$. All of the other variables are dependent on the pressure, and in addition $(\gamma_i)_L$ is explicitly dependent on the liquid composition x_i. A numerical procedure that includes an iterative solution of the type 1 bubble-point calculation is employed:

1. From the given temperature and vapor composition calculate the other constants of the problem (vapor pressures, liquid molal volumes, vapor activity coefficients, and fugacity ratios of the pure components at their vapor pressures).
2. Assume a liquid composition x_i.
3. Do a type 1 bubble-point calculation using the assumed x_i.
4. Test the calculated vapor mole fraction against the given values.
5. If the agreement is unsatisfactory, repeat the calculation from step 3, using values of x_i calculated from Equation 5-65 as the new assumed values.

The following example illustrates the use of the above procedure:

Example 5-14

Use the data of Example 5-10 to calculate the dew-point pressure at 50°C and the corresponding liquid composition for a mixture of methyl acetate, acetone, and methanol having mole fractions of 0.33, 0.34, and 0.33, respectively.

Solution. A FORTRAN program for the solution of the problem is shown in Figure 5.42. This program is identical with the one used in Example 5-10 through line 112.

Lines 112 through 127 set up the values of liquid mole fractions to be used in the next iteration.

Lines 112 through 114 direct the computation of the sum of the squares of the errors in the calculated vapor mole fraction.

Line 116 tests this sum for convergence.

Lines 122 through 124 calculate new values for the liquid mole fractions according to the equation

$$x_i = y_i \left(\frac{x_i}{y_i}\right)_{\text{calc}}.$$

Lines 126 and 127 normalize the values of the x terms to make them add up to 1.

Line 131 directs the initiation of the next iteration.

Figure 5.43 shows the computer output of the solution to this problem.

The second type of dew-point calculation is similar to the type 2 bubble-point calculation:

Given a vapor of composition $y_1, y_2, \ldots, y_n$, at a known total pressure, calculate the temperature at which liquid will begin to condense and the composition of the condensing liquid.

The constants of the type 2 dew-point calculation are total pressure π and vapor composition y_i. The values of $(\gamma_i)_L$, $(\gamma_i)_G$, p_i^*, η_i^*, η_i^π, and $(v_i)_L$ are all

```
0    $IBFTC E1083Y
1         DIMENSION G(3,3),A(3,3),ANT(3,3),P(3),X(3),Y(3),GAM(3),V(3),
        1 SUMA(3),YG(3)
2     1 READ(5,1000) T
3       READ(5,1000) ANT
4       READ(5,1000) G
5       READ(5,1000) V
6       READ (5,1000) YG
7       DO 3 I=1,3
10         A(I,I) = 1.
11      DO 3 J=1,3
12      IF(J.LE.I) GO TO 3
15         B1=-ALOG(G(I,J))
16         B2=-ALOG(G(J,I))
17    2    F1 = ALOG(G(I,J)) + B1 + EXP(B2) - 1.
20         F2 = ALOG(G(J,I)) + B2 + EXP(B1) - 1.
21         DB1=F1-F2*EXP(B2)
22         DB2=F2-F1*EXP(B1)
23      IF(ABS(B1+B2).LT.0.05) GO TO 22
26         DEN=EXP(B1+B2)-1.
27         DB1=DB1/DEN
30         DB2=DB2/DEN
31   22    B1 = B1 + DB1
32         B2 = B2 + DB2
33      IF (ABS(DB1)+ABS(DB2).GT.0.00001) GO TO 2
36         A(I,J) = EXP(B1)
37         A(J,I) = EXP(B2)
40    3 CONTINUE
43      WRITE (6,1001) A
44      WRITE (6,1002) G
45      WRITE (6,1009) V
46      DO 4 I=1,3
47    4    P(I)=10.**(ANT(1,I) - ANT(2,I)/(T+ ANT(3,I)))
51      WRITE (6,1005) P
52  501 READ (5,1000) X
53  502 WRITE (6,1003) X
54      DO 5 I = 1,3
55         SUMA(I) = 0.
56      DO 5 J = 1,3
57    5    SUMA(I) = SUMA(I) + A(I,J)*X(J)
62      DO 7 I = 1,3
63         ALNG = 1. - ALOG(SUMA(I))
64      DO 6 J = 1,3
65    6    ALNG = ALNG - A(J,I)*X(J)/SUMA(J)
67    7    GAM(I) = EXP(ALNG)
71      WRITE (6,1008) GAM
72         PI = P(1)*X(1) +P(2)*X(2) +P(3)*X(3)
73      WRITE (6,1010) PI
74   11    PIC = 0.
75      DO 12 I = 1,3
76   12    PIC = PIC + GAM(I)*P(I)*X(I)*EXP(V(I)*(PI-P(I))/(62370.*
        1    (T+ 273.16)))
100     WRITE (6,1010) PIC
101     IF (ABS((PIC-PI)/PI).LT.(0.1**5)) GO TO 13
104        PI = PIC
```

Figure 5.42 FORTRAN program for the solution of Example 5-14.

```
105            GO TO 11
106         13 DO 14 I = 1,3
107         14     Y(I) = GAM(I)*P(I)*X(I)*EXP(V(I)*(PI-P(I))/(62370.*
             1            (T +273.16)))/PI
111            WRITE (6,1011) Y
112                SUM    = 0.
113            DO 30 I = 1,3
114         30     SUM    = SUM + (Y(I)-YG(I))**2
116            IF (SUM .LT.(0.1**8)) GO TO 1
121                SUM    = 0.
122            DO 31 I = 1,3
123                X(I)   = YG(I)*X(I)/Y(I)
124         31     SUM    = SUM + X(I)
126            DO 32 I = 1,3
127         32     X(I)   = X(I)/SUM
131            GO TO 502
132       1000 FORMAT (G0.0)
133       1001 FORMAT ( 18H1 WILSON CONSTANTS/9F12.5)
134       1002 FORMAT ( 18H0 GAM(INF. DIL.)  /9F12.5)
135       1003 FORMAT ( 12H0 LIQ. M. F./3F12.5)
136       1005 FORMAT ( 18H0 VAPOR PRESSURES /3F12.5)
137       1008 FORMAT ( 24H0 ACTIVITY COEFFICIENTS /3F12.6)
140       1009 FORMAT ( 18H0 LIQUID VOLUMES  /3F12.6)
141       1010 FORMAT ( 12H0 PRESSURE = F12.4)
142       1011 FORMAT ( 12H0 VAPOR M.F. /3F12.6)
143            END
```

Figure 5.42 (*Continued*).

dependent on temperature. In addition, $(\gamma_i)_L$ is dependent on the liquid composition. The following numerical solution procedure is quite similar to the type 2 bubble-point calculation:

1. Assume a value for the temperature T.
2. Do a type 1 calculation for π and x_i.
3. Test the calculated total pressure against the given total pressure.
4. If the calculated pressure is less than the given pressure, increase the assumed temperature; if the calculated pressure is greater than the given pressure, decrease the assumed temperature. Repeat from step 2 until the calculated pressure agrees sufficiently well with the given pressure. As an alternative method, the pressure may be plotted against temperature and the solution read from the graphs.

This procedure is illustrated in the following example:

Example 5-15

Use the data of Example 5-10 to calculate the dew-point temperature at atmospheric pressure (760 mm Hg) for a mixture of methyl acetate, acetone, and methanol having mole fractions of 0.33, 0.34, and 0.33, respectively. Assume that $\ln \gamma^\infty$ varies inversely with absolute temperature.

Solution. The solution is obtained by repeating the calculation of Example 5-14 at a number of temperatures. The values of γ^∞ for input are adjusted for each temperature by

$$\ln \gamma^\infty = \ln \gamma^\infty \Big|_{50} \left(\frac{323}{t + 273}\right).$$

LIQ. M. F.
0.33000   0.34000   0.33000

ACTIVITY COEFFICIENTS
1.201301   1.029787   1.418105

PRESSURE =   539.4190

PRESSURE =   641.9891

PRESSURE =   642.2138

PRESSURE =   642.2143

VAPOR M.F.
0.361458   0.334613   0.303930

LIQ. M. F.
0.29976   0.34373   0.35650

ACTIVITY COEFFICIENTS
1.228281   1.038220   1.378264

PRESSURE =   535.0582

PRESSURE =   639.3398

PRESSURE =   639.5648

PRESSURE =   639.5653

VAPOR M.F.
0.337099   0.342467   0.320434

LIQ. M. F.
0.29291   0.34063   0.36647

ACTIVITY COEFFICIENTS
1.237582   1.041646   1.365501

PRESSURE =   533.2912

PRESSURE =   638.5144

PRESSURE =   638.7403

PRESSURE =   638.7408

VAPOR M.F.
0.332311   0.340929   0.326760

LIQ. M. F.
0.29068   0.33947   0.36985

ACTIVITY COEFFICIENTS
1.240773   1.042852   1.361261

PRESSURE =   532.6878

PRESSURE =   638.2271

PRESSURE =   638.4534

Figure 5.43 Solution for Example 5-14.

PRESSURE =   638.4538

VAPOR M.F.
0.330778   0.340319   0.328903

LIQ. M. F.
0.28992   0.33908   0.37100

ACTIVITY COEFFICIENTS
1.241859   1.043266   1.359834

PRESSURE =   532.4832

PRESSURE =   638.1286

PRESSURE =   638.3549

PRESSURE =   638.3554

VAPOR M.F.
0.330264   0.340109   0.329629

LIQ. M. F.
0.28967   0.33894   0.37139

ACTIVITY COEFFICIENTS
1.242228   1.043407   1.359351

PRESSURE =   532.4138

PRESSURE =   638.0951

PRESSURE =   638.3214

PRESSURE =   638.3219

VAPOR M.F.
0.330089   0.340037   0.329874

LIQ. M. F.
0.28958   0.33889   0.37152

ACTIVITY COEFFICIENTS
1.242354   1.043455   1.359187

PRESSURE =   532.3903

PRESSURE =   638.0837

PRESSURE =   638.3100

PRESSURE =   638.3105

VAPOR M.F.
0.330030   0.340013   0.329958

Figure 5.43 (*Continued*).

5.5 Process Calculations

The results of several calculations are shown in Table 5.18.

Table 5.18 Results of Calculation for Example 5-15

$T(°C)$	Pressure (mm Hg)	Liquid-phase Mole Fraction		
		Methyl Acetate	Acetone	Methanol
50	638.3	0.2896	0.3389	0.3715
52	688.7	0.2918	0.3413	0.3669
55	770.4	0.2953	0.3449	0.3598
60	923.5	0.3002	0.3503	0.3495

These results shown in Table 5.18 are plotted in Figure 5.44, from which a temperature of 54.5°C is seen to correspond to 760 mm Hg total pressure. The corresponding liquid-phase mole fractions are as follows:

Methyl acetate	0.2947
Acetone	0.3445
Methanol	0.3608

5.5.3 Flash Calculations

The flash-calculation problem can be generally stated as follows:

Given a mixture whose bulk composition is $X_1, X_2, \ldots, X_n$, existing at total pressure π and temperature T, determine the fraction of the mixture that will be in the vapor phase at equilibrium and the compositions of the liquid and vapor phases.

The solution of the flash problem combines a material balance with the phase-equilibrium calculation. If we let the Greek letter ψ represent the fraction of the total mixture in the vapor phase at equilibrium, a material balance on component i gives

$$X_i = \psi y_i + (1 - \psi)x_i. \tag{5-66}$$

Phase-equilibrium relationships give

$$y_i = K_i x_i,$$

where

$$K_i = \frac{(\gamma_i)_L \eta_i^* p^*}{(\gamma_i)_G \eta_i^\pi \pi} \exp\left[\frac{(v_i)_L(\pi - p_i^*)}{RT}\right]. \tag{5-67}$$

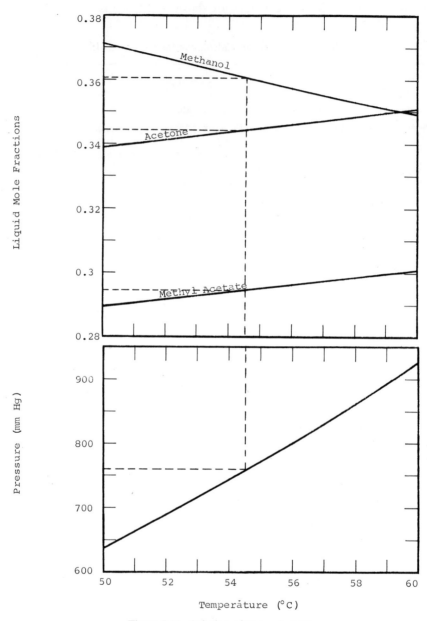

Figure 5.44 Solution of Example 5-15.

5.5 Process Calculations

The preceding equations can be combined to eliminate y_i and solved for x_i:

$$x_i = \frac{X_i}{K_i \psi + 1 - \psi}. \qquad (5\text{-}68)$$

The condition that the sum of all liquid-phase mole fractions must equal unity can be used to define a function w, which must be zero:

$$w = \sum_i x_i - 1 = 0. \qquad (5\text{-}69)$$

The solution of Equation 5-69 can usually be accomplished by a Newton–Raphson method using ψ as the independent variable. In order to use the Newton–Raphson method the derivative of w is expressed as

$$w' = \sum_i \frac{dx_i}{d\psi} \simeq -\sum_i \frac{X_i(K_i - 1)}{(K_i \psi + 1 - \psi)^2}. \qquad (5\text{-}70)$$

Equations 5-66 through 5-70 form the basis of the flash calculation. However, it is wise to do two preliminary calculations before the flash calculation is attempted. First, a type 1 bubble-point calculation should be made. If the value of π determined by this calculation is less than the given value of π, the given temperature is below the bubble-point temperature and the entire mixture is a supercooled liquid ($x_i = X_i$); no further calculation need be made. If the mixture is not a supercooled liquid, a type 1 dew-point calculation should be made. If the calculated pressure is greater than the given pressure, the temperature is above the dew-point temperature and the entire mixture is a superheated vapor ($y_i = X_i$); no further calculation need be made. If the mixture is neither a superheated vapor nor a supercooled liquid, a flash calculation can be carried out by the following procedure:

1. Assume a value for ψ and liquid composition x_i.
2. Calculate the values of y_i by Equation 5-67, assuming unity for the vapor-phase activity coefficient.
3. Calculate liquid-phase activity coefficients from the values of x_i and vapor-phase activity coefficients from the values of y_i.
4. Calculate the values of K_i by Equation 5-67.
5. Calculate new values of the x_i terms by Equation 5-68.
6. Calculate w by Equation 5-69; if w is sufficiently near zero, skip to step 10.
7. Calculate w' according to Equation 5.70.
8. Increment ψ by $\psi = \psi - w/w'$, provided ψ remains between 0 and 1.
9. Repeat from step 5, using new values of ψ and x_i.
10. Test the values of x_i and y_i to see if they have changed significantly between steps 2 and 5; if there is significant change, go back to step 2.

11. If the x values have not changed, the current values for y_i, ψ, and x_i are the correct solution to the problem.

This procedure is illustrated in the following example:

Example 5-16

Perform a flash calculation at 50°C and 640 mm Hg for a feed solution of 0.33, 0.34, and 0.33 mole fraction methyl acetate, acetone, and methanol, respectively. Assume the vapor phase to be ideal throughout; use the Wilson equations for liquid-phase activity coefficients and use the data of Example 5-10 for infinite-dilution activity coefficients. Calculate the compositions of the resulting liquid and vapor phases, and the fraction of the feed vaporized.

Solution. Bubble-point and dew-point calculations for this mixture were made in Examples 5-10 and 5-14, respectively. These calculations show that the given pressure (640 mm Hg) lies between the bubble-point pressure and the dew-point pressure at 50°C. A FORTRAN program for the solution of the flash equation is shown in Figure 5.45, and the solution obtained by using the program is shown in Figure 5.46. Since the program follows quite closely the calculation procedure described, a detailed explanation of the FORTRAN listing is not given for this problem.

5.5.4 Complete Vapor–Liquid Equilibrium Phase Diagram

If a complete phase diagram is required for a vapor–liquid system, the range of pressures and temperatures and of liquid compositions over which the diagram is desired must be determined. Type 1 bubble-point calculations are then made at each of a sufficient number of points to define the phase diagram adequately. The results are then plotted or cross-plotted (or both) to yield the type of diagram desired.

5.5.5 Process Applications

The major area of application of vapor–liquid equilibrium calculations is in the analysis and the design of separation equipment. Some of the unit operations involved in such calculations are evaporation, distillation, and gas absorption. Design calculations are usually based on either the ideal-stage concept or mass-transfer theory applied to the two-film concept. Both of these require knowledge of vapor–liquid equilibrium.

Ideal-Stage Calculations. The ideal stage as used in vapor–liquid contacting operations is shown schematically in Figure 5.47. Such stages most frequently take the form of plates in a vertical column. There may or may not be an external feed to any given ideal stage, as indicated by stream F. There will be

```
   0     $IBFTC E1083U
   1           DIMENSION XF(3),X(3),Y(3),A(3,3),P(3),TITLE(12),V(3),K(3),AK(3),
               1          GAM(3),SUM(3),XC(3)
   2           REAL    K
   3     1     KEY     = 0
   4           KEY2    = 0
   5           READ    (5,1001) TITLE
   6           READ    (5,1000) A
   7           READ    (5,1000) T,PI,R,P,V
  10           READ    (5,1000) XF
  11           READ    (5,1000) PSI,X
  12           WRITE   (6,1002) TITLE
  13           WRITE   (6,1003) A
  14           WRITE   (6,1004) T,PI,R,P,V
  15           WRITE   (6,1005) XF
  16           T       = T + 273.16
  17           DO 10 I = 1,3
  20     10    AK(I)   = P(I)*EXP((PI-P(I))/(R*T))
         C
  22     20    DO 21 I = 1,3
  23           SUM(I)  = 0.
  24           DO 21 J = 1,3
  25     21    SUM(I)  = SUM(I) + A(I,J)*X(J)
  30           DO 23 I = 1,3
  31           ALNG    = 1. - ALOG(SUM(I))
  32           DO 22 J = 1,3
  33     22    ALNG    = ALNG - A(J,I)*X(J)/SUM(J)
  35           GAM(I)  = EXP(ALNG)
  36     23    K(I)    = GAM(I)*AK(I)/PI
  40    231    W       = -1.
  41           KEY2    = KEY2 + 1
  42           IF ( KEY2 .GT. 25) GO TO 1
  45           WPR     = 0.
  46           SUMX    = 0.
  47           DO 232 I = 1,3
  50           XC(I)   = XF(I)/((K(I)-1.)*PSI+1.)
  51           Y(I)    = K(I)*XC(I)
  52           W       = W+XC(I)
  53           WPR     = WPR-(K(I)-1.)*XF(I)/((K(I)-1.)*PSI+1.)**2
  54    232    SUMX    = SUMX + (X(I)-XC(I))**2
  56           IF(ABS(W).LT. 0.1**6) KEY = 1
  61           WRITE (6,1006) PSI
  62           WRITE (6,1007) XC
  63           WRITE (6,1008) Y
  64           IF (KEY .EQ. 1) GO TO 27
  67     25    B       = PSI - W/WPR
  70           IF (B .LE. 1. .AND. B .GE. 0.) GO TO 26
  73           W       = 0.5*W
  74           GO TO 25
  75     26    PSI     = B
  76           GO TO 231
  77     27    IF(SUMX .LE. 0.1**8) GO TO 1
 102           DO 28 I = 1,3
 103     28    X(I)    = XC(I)
 105           GO TO 20
 106    1000   FORMAT (G0.0)
 107    1001   FORMAT (12A6)
 110    1002   FORMAT (1H1,12A6)
 111    1003   FORMAT ( 27H0  WILSON EQUATION CONSTANTS/3X,3G15.5/3X,3G15.5/
               1          3X,3G15.5)
 112    1004   FORMAT ( 18H0 TEMPERATURE   =   G12.4,    12H PRESSURE = G12.4,
               1          18H GAS CONSTANT   =   G12.4/    18H VAPOR PRESSURES  /
               2          3G12.4, /    18H LIQUID VOLUMES    / 3G12.4)
 113    1005   FORMAT ( 18H0 FEED COMPOSITION/ 3G15.5)
 114    1006   FORMAT ( ///     18H FRACTION VAPOR   =   G12.5)
 115    1007   FORMAT ( 24H LIQUID COMPOSITION       / 3G15.5)
 116    1008   FORMAT ( 24H VAPOR   COMPOSITION      / 3G15.5)
 117           END
```

Figure 5.45 FORTRAN program for solution of Example 5-16.

FLASH CALCULATION – METHYL ACETATE, ACETONE, METHANOL AT 50 DEG. C.

WILSON EQUATION CONSTANTS

1.00000	0.71891	0.57939
1.18160	1.00000	0.97513
0.52297	0.50878	1.00000

TEMPERATURE = 50.000   PRESSURE = 640.00

VAPOR PRESSURES

585.42	613.69	416.89

LIQUID VOLUMES

83.770	76.810	42.050

FEED COMPOSITION

0.33000	0.34000	0.33000

FRACTION VAPOR = 0.50000
LIQUID COMPOSITION

0.31446	0.34215	0.34308

VAPOR COMPOSITION

0.34554	0.33785	0.31692

FRACTION VAPOR = 0.66698
LIQUID COMPOSITION

0.30959	0.34287	0.34768

VAPOR COMPOSITION

0.34019	0.33857	0.32117

FRACTION VAPOR = 0.62791
LIQUID COMPOSITION

0.31071	0.34270	0.34659

VAPOR COMPOSITION

0.34143	0.33840	0.32017

FRACTION VAPOR = 0.62551
LIQUID COMPOSITION

0.31078	0.34269	0.34653

VAPOR COMPOSITION

0.34151	0.33839	0.32010

FRACTION VAPOR = 0.62551
LIQUID COMPOSITION

0.30801	0.34164	0.35032

VAPOR COMPOSITION

0.34316	0.33902	0.31783

FRACTION VAPOR = 0.62551
LIQUID COMPOSITION

0.30744	0.34139	0.35109

VAPOR COMPOSITION

0.34350	0.33917	0.31737

Figure 5.46 Solution of Example 5-16.

```
FRACTION VAPOR   =  0.62551
LIQUID COMPOSITION
     0.30732            0.34133          0.35124
VAPOR   COMPOSITION
     0.34358            0.33921          0.31729


FRACTION VAPOR   =  0.62551
LIQUID COMPOSITION
     0.30728            0.34131          0.35126
VAPOR   COMPOSITION
     0.34360            0.33922          0.31727
```

Figure 5.46 (*Continued*).

flowing into the ideal stage a vapor rising from the stage below or from an external source, indicated by stream V_{i-1}. There will also be a liquid stream to this stage coming from the overflow of a higher stage or from some external source. This liquid feed is indicated by stream L_{i+1}. Leaving each stage will be a vapor stream V_i and a liquid stream L_i. There may be withdrawn from each of these stages a product sidestream indicated by V_{ss} for the vapor stream or L_{ss} for the liquid stream.

Analysis of such an ideal stage consists simply of a flash calculation where the total mixture consists of the stream $F + V_{i-1} + L_{i+1}$. The liquid resulting

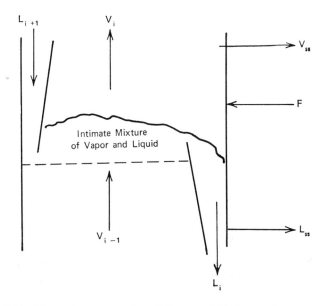

Figure 5.47 Schematic representation of the general ideal stage for vapor–liquid phase contacting.

from the flash calculation consists of the sum of streams L_{ss} and L_i. The vapor stream of the flash calculation consists of streams $V_{ss} + V_i$. The quantities of the product streams L_{ss} and V_{ss} must be known either as absolute quantities or as fractions of the total streams from which they come.

The complete design of a column consists of computing the interactions of a number of such stages, combined with overall segmental material and energy balances. The complete calculation of the entire separation process is beyond the scope of this work but is covered in a number of good texts, particularly, the work of Smith [59]. This illustration is included merely to indicate the direct and intimate relationship between phase-equilibrium principles and process-equipment-design calculations.

Mass-Transfer Calculations. Mass-transfer calculations are usually based on some form of two-film theory relating the rate of transfer of material from one phase to the other to the departure from equilibrium between the phases. If it is expressed in terms of an overall liquid-phase coefficient, the rate of mass transfer is given by

$$\frac{dN_i}{dt} = K_{OL} A (x_i - x_i^*),$$

where dN_i/dt = rate of transfer of component i from liquid phase to vapor phase,
K_{OL} = overall mass transfer coefficient based on liquid composition,
A = interfacial area between phases,
x_i = the liquid-phase composition,
x_i^* = the composition the liquid phase would have if it were at equilibrium with the vapor-phase composition y_i.

The determination of x_i^* from y_i represents a dew-point calculation.

The calculations may also be made with vapor-based overall mass-transfer coefficients:

$$\frac{dN_i}{dt} = K_{OG} A (y_i^* - y_i),$$

where K_{OG} is the overall mass-transfer coefficient based on gas-phase composition and y_i^* is the composition the liquid phase would have if it were at equilibrium with the liquid-phase composition x_i.

The determination of y_i^* from x_i represents a bubble-point calculation.

In terms of individual films the rate of transfer is given by

$$\frac{dN_i}{dt} = k_L A (x_i - x_i^I) = k_G A (y_i^I - y_i),$$

where k_L and k_G are the mass-transfer coefficients for the liquid and vapor films, respectively, and x_i^I and y_i^I are the concentrations of liquid and vapor phases at the phase interface.

The compositions x_i^I and y_i^I are usually assumed to be at equilibrium, and their values are obtained from simultaneous solution of the two equations

$$y_i^I = y_i + \frac{k_L}{k_G}(x_i - x_i^I),$$

$$y_i^I = K_i x_i^I.$$

In the design of a complete phase-contacting column these calculations must be made at numerous points in the column and combined with the material and energy balances that apply.

REFERENCES

[1] L. H. Horsley, *Azeotropic Data*, American Chemical Society, Washington, D.C., 1952.
[2] W. B. Kay, *Ind. Eng. Chem.*, **30**, 459 (1938).
[3] E. Hala, J. Pick, V. Fried, and O. Vilim, *Vapor-Liquid Equilibrium*, 2nd ed., Pergamon Press, London, 1968.
[4] F. C. Collins and V. Lantz, *Ind. Eng. Chem., Anal. Ed.*, **18**, 673 (1946).
[5] S. R. M. Ellis and R. M. Contractor, *J. Inst. Petr.*, **45**, 147 (May 1959).
[6] R. H. McCormick, P. Barton, and M. R. Fenske, *Am. Inst. Chem. Engrs. J.*, **8**, 365 (1962).
[7] C. F. Oldershaw, *Ind. Eng. Chem., Anal. Ed.*, 265 (1941).
[8] R. W. Sivanson and J. A. Gerster, *J. Chem. Engr. Data*, **7**, 132 (1962).
[9] E. R. Adland, M. A. Khan, and B. T. Whitman, *Gas Chromatography*, Butterworths, London, 1961, p. 251.
[10] C. F. Chueh, Ph.D. Thesis, Georgia Institute of Technology, July 1962.
[11] K. H. Everett and C. T. H. Stoddart, *Trans. Faraday Soc.*, **57**, 746 (1961).
[12] C. J. Hardy, *J. Chromatog.*, **2**, 490 (1959).
[13] A. I. M. Keulmans, *Gas Chromatography*, Reinhold, New York, 1959, pp. 106 and 182.
[14] A. Kivante and W. A. Rijnders, *Gas Chromatography*, Reinhold, New York, 1955, pp. 125 ff.
[15] G. L. Mellado and R. Kobayaski, *Petr. Refiner*, 125, (February 1960).
[16] G. J. Pierotti, C. H. Deal, E. L. Derr, and P. E. Porter, *J. Am. Chem. Soc.*, **78**, 2989 (1955).
[17] P. E. Porter, C. H. Deal, and F. H. Stross, *J. Am. Chem. Soc.*, **78**, 2999 (1956).
[18] G. W. Warren, R. R. Warren, and V. A. Yarbrough, *Ind. Eng. Chem.*, **51**, 1475 (1959).
[19] R. W. Dornte, *J. Phys. Chem.*, **33**, 1309 (1929).
[20] M. S. Wrewski, *J. Russ. Phys. Chem. Soc.*, **42**, 1 (1910).
[21] A. Wilson and E. L. Simms, *Ind. Eng. Chem.*, **44**, 2214 (1952).
[22] R. C. Reid and T. K. Sherwood, *The Properties of Gases and Liquids*, McGraw-Hill Book Company, New York, 1958, pp. 321–323.

[23] R. H. Fariss and V. J. Law, "Practical Tactics for Overcoming Difficulties in Non-Linear Regression and Equation Solving," paper presented at 61st National Meeting of the American Institute of Chemical Engineers, February 1967.
[24] M. A. Rosanoff, C. W. Bacon, and J. F. Schulze, *J. Am. Chem. Soc.*, **36**, 1993 (1914).
[25] Monsanto Company data.
[26] A. R. Martin and B. Collie, *J. Chem. Soc. (London)*, 2658 (1932).
[27] B. H. Sage and W. N. Lacey, *Ind. Eng. Chem.*, **32**, 992 (1940).
[28] C. P. Smyth and E. W. Engel, *J. Am. Chem. Soc.*, **51**, 2646 (1929).
[29] I. Brown and A. H. Ewald, *Australian J. Sci. Research*, **3A**, 306 (1950).
[30] J. Griswold and S. Y. Wong, *Chem. Eng. Progr. Symposium Series*, No. 3, **48**, 18 (1952).
[31] W. G. Beare, G. A. McVicar, and J. B. Teague, *J. Phys. Chem.*, **34**, 1310 (1930).
[32] A. R. Martin and C. M. George, *J. Chem. Soc., (London)*, 1413 (1933).
[33] P. C. Teague and W. A. Felsing, *J. Am. Chem. Soc.*, **65**, 485 (1943).
[34] I. Brown and A. H. Ewald, *Australian J. Sci. Research*, **4A**, 198 (1951).
[35] G. A. Stein, B.S. Thesis, Massachusetts Institute of Technology, 1937.
[36] J. Someshima, *J. Am. Chem. Soc.*, **40**, 1503 (1918).
[37] J. Lewin, *Bull. Soc. Chim. Belg.*, **39**, 91 (1930).
[38] C. P. Smyth and E. W. Engel, *J. Am. Chem. Soc.*, **51**, 2660 (1929).
[39] J. R. Lacher, W. B. Buck, and W. H. Parry, *J. Am. Chem. Soc.*, **63**, 2422 (1941).
[40] J. Hirschberg, *Bull. Soc. Chim. Belg.*, **41**, 169 (1932).
[41] A. R. Gordon and W. J. Hornibrook, *Can. J. Research*, **24B**, 263 (1946).
[42] J. H. Ashley and G. R. Brown, *Chem. Eng. Progr. Symposium Series*, No. 10, **50**, 129 (1954).
[43] H. H. Reamer, B. H. Sage, and W. N. Lacey, *Ind. Eng. Chem.*, **44**, 602 (1952).
[44] C. E. Funk, Jr., *Ind. Eng. Chem.*, **43**, 1152 (1951).
[45] F. Ishikawa and T. Yamaguchi, *Bull. Inst. Chem. Res. (Tokyo)*, **17**, 246 (1938).
[46] J. A. Barken, I. Brown, and F. Smith, *Discussions Faraday Soc.*, No. 15, 142 (1953).
[47] J. B. Ferguson, M. Freed, and A. C. Morris, *J. Phys. Chem.*, **37**, 87 (1933).
[48] C. B. Kretschmer, J. Nowakowska, and R. Wiebe, *J. Am. Chem. Soc.*, **70**, 1785 (1948).
[49] C. B. Kretschmer and R. Wiebe, *J. Am. Chem. Soc.*, **71**, 1793 (1949).
[50] W. A. Wright, *J. Phys. Chem.*, **37**, 233 (1933).
[51] J. B. Ferguson, *J. Phys. Chem.*, **36**, 1123 (1932).
[52] S. C. Lee, *J. Phys. Chem.*, **35**, 3558 (1931).
[53] C. R. Fordyce and D. R. Simonson, *Ind. Eng. Chem.*, **41**, 104 (1949).
[54] A. L. Vierk, *Z. Anorg. Chem.*, **261**, 283 (1950).
[55] J. Griswold, R. V. West, and K. K. McMillin, *Chem. Eng. Progr. Symposium Series*, No. 2, **48**, 62 (1952).
[56] M. F. Gautreaux and J. Coates, *Am. Inst. Chem. Engrs. J.*, **1**, 496 (1955).
[57] W. H. Severns, A. Sesonski, R. H. Perry, and R. L. Pigford, *Am. Inst. Chem. Engrs. J.*, **1**, 401 (1955).
[58] C. Black, *Ind. Eng. Chem.*, **50**, 391 (1958).
[59] B. D. Smith, *Design of Equilibrium Stage Processes*, McGraw-Hill Book Company, New York, 1963.

PROBLEMS

1. Test the following data of Griswold and Wong [30] for thermodynamic consistency. The system is acetone–methanol at 100°C.

Problems

Mole Fraction Acetone in		Pressure (mm Hg)
Liquid	Vapor	
0.047	0.07	2643
0.068	0.093	2674
0.146	0.187	2818
0.22	0.258	2844
0.397	0.417	3010
0.507	0.507	3020
0.562	0.547	3020
0.624	0.598	3020
0.641	0.614	3010
0.660	0.635	3015
0.747	0.711	2968
0.87	0.819	2891
0.952	0.917	2813
0.977	0.95	2762

In making the test, assume (a) ideal gas phase; (b) ideal solution in gas phase, but Redlich–Kwong equation of state for pure components; (c) gas-phase mixtures obey the Redlich–Kwong equation of state. Choose your own mixing rule.

2. Test the following data for the nitrogen–oxygen system [E. C. C. Baly, *Proc. Phys. Soc.* (*London*), **17**, 1957 (1900)] for thermodynamic consistency. The pressure is 1 atmosphere throughout.

Mole Fraction Nitrogen		T (°K)
Liquid	Vapor	
0.059	0.1545	89.5
0.1018	0.2563	88.5
0.1469	0.3515	87.5
0.1956	0.437	86.5
0.2490	0.5193	85.5
0.3069	0.5955	84.5
0.3707	0 6665	83.5
0.4406	0.7327	82.5
0.5208	0.7878	81.5
0.6147	0.8522	80
0.7233	0.9067	79
0.919	0.9782	78

3. Determine Van Laar and Wilson equation parameters for the data of Problem 1. The use of a computer is recommended for this problem.

4. Perform the regression analysis indicated in the experimental data in Table 5.10.

5. In a Oldershaw column with an efficiency of 0.6 the following data are obtained in two successive runs on a binary system:

	Mole Fraction Component 1		
Run No.	Bottoms	Overhead	Number of Stages
1	0.10	0.49	10
2	0.11	0.50	20

Calculate the relative volatility in each run and comment on the significance of the results.

6. Determine the infinite-dilution activity coefficients from the pressure–liquid composition data of Problem 1. Use these values in the Van Laar equation to calculate vapor mole fractions; compare calculated and experimental values.

7. Repeat Problem 6 using temperature–liquid composition data of Problem 2 and the Wilson equations.

8. The following values of infinite-dilution activity coefficients have been found at 50°C:

Binary	$\ln \gamma_1^\infty$	$\ln \gamma_2^\infty$
Butyl bromide–n-heptane	0.1581	0.1649
Butyl chloride–butyl bromide	0	0
Butyl chloride–n-heptane	0.1071	0.2815

Calculate, using the equation of your choice, the atmospheric bubble point and dew point of an equimolar mixture of butyl bromide, butyl chloride, and n-heptane. Assume $\ln \gamma^\infty \propto 1/T$. Also calculate liquid and vapor fractions and compositions for an atmospheric flash at a temperature midway between the bubble point and the dew point.

Chapter 6

LIQUID–LIQUID EQUILIBRIUM

6.1 INTRODUCTION

The possible existence of two or more liquid phases simultaneously must frequently be considered in process applications of phase equilibrium. If limited miscibility occurs unexpectedly, it can be a serious disadvantage; for example, two-phase pumping is more difficult than single-phase pumping, and capacity and efficiency are both adversely affected in distillation columns when two liquid phases exist internally. When two liquid phases occur in process flow streams, extra processing steps are required to either keep them well mixed if separation is not desired or to allow them to settle when phase separation is desired.

If limited liquid-phase miscibility is expected, however, advantage can be taken of it in process design. Liquid–liquid extraction is a powerful separation method based entirely on limited liquid miscibility and the high selectivity of the phases between various components. In distillation the overhead product may have limited miscibility after condensation. The liquid-phase separation, properly considered in design, then gives an additional highly selective stage of separation. If azeotropic distillation is used, an additional component not present in the feed stream is added to enhance the relative volatility of the key components. Since this extra component (called the entrainer) must subsequently be separated from the product, limited liquid miscibility is highly desirable.

The mathematical treatment of liquid–liquid equilibrium is, in many ways, more complex than that of vapor–liquid equilibrium. The primary reason for the difficulty is that neither phase is ever an ideal solution. Indeed, if an ideal solution exists in the liquid phase, there can never be more than a single liquid phase formed.

6.2 TYPES OF PHASE DIAGRAM

Two types of phase diagram frequently appear in the literature. Binary systems are generally represented by constant-pressure plots having temperature as the ordinate and composition (in terms of the mole fraction of one component) as the abscissa. Ternary (three-component) systems are usually plotted on triangular diagrams for a specific temperature and a specific pressure. Each vertex of the triangle represents one of the pure components, with all real compositions as points on the boundary or interior of the triangle. There is no generally accepted method of representing systems of more than three components graphically.

6.2.1 Binary Systems

The most general type of phase diagram for a binary system is shown in Figure 6.1. At temperatures higher than the upper critical solution temperature (designated c.s.t.) the two components are miscible in all proportions; thus there is only one liquid phase. The same condition exists for temperatures below the lower critical solution temperature. At temperatures between the two c.s.t. points, however, the two components are not completely soluble.

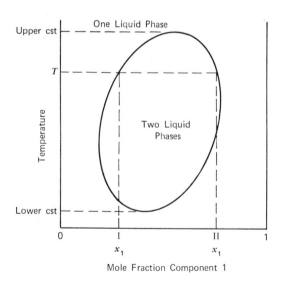

Figure 6.1 Typical liquid–liquid phase diagram for a binary system at constant pressure.

6.2 Types of Phase Diagram

At the intermediate temperature T in the diagram only one phase will be formed if the overall composition of a mixture is less than x_1^I mole fraction of component 1 or greater than x_1^{II}. At intermediate overall compositions, however, the mixture will form two phases, one having composition x_1^I and the other x_1^{II}. The relative quantities of the two phases is determined by a material balance and the overall composition.

The diagram in Figure 6.1 assumes that the temperature is below the bubble point of all compositions at the pressure of the diagram. If the temperature scale is extended vertically there will ultimately be a temperature at which vaporization occurs. Thus a vapor–liquid phase diagram will appear *above* the liquid–liquid diagram, as shown in Figure 6.2. If the upper c.s.t. is

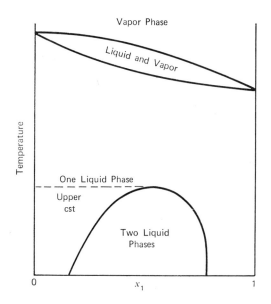

Figure 6.2 Combined liquid–liquid and vapor–liquid binary phase diagram.

sufficiently high for the two-phase region to intersect the bubble-point curve, a diagram of the type depicted in Figure 5.5 or 5.6 results.

Not all of the systems that exhibit limited miscibility have a lower critical solution temperature. For those that do not the two-phase region ultimately intersects the freezing-point curve at lower temperatures and solid phases appear. A sample diagram depicting limited miscibility throughout the entire liquid-phase temperature span is shown in Figure 6.3.

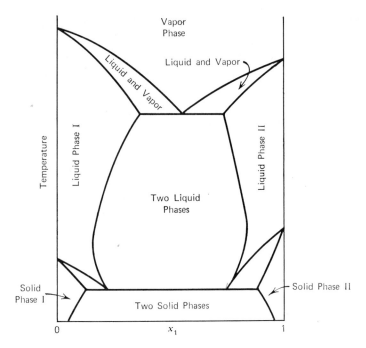

Figure 6.3 Complete binary phase diagram for partially miscible system with no critical solution temperature.

6.2.2 Ternary Systems

Phase diagrams for ternary liquid systems are usually shown at constant temperature and pressure on triangular diagrams. On an equilateral triangle each vertex represents one of the pure components. Figure 6.4 shows how compositions are represented on such diagrams. Two coordinates are sufficient to determine a composition on the diagram, since the third mole fraction can be obtained by subtraction. Lines of constant mole fraction of any component are parallel to the side opposite the vertex representing the pure component. The sides of the triangle represent binary systems, since the mole fraction of the component at the opposite vertex is zero. Points on the interior of the triangle represent three components.

If all components are completely miscible in all proportions, the entire diagram is a single phase; hence the triangle is blank, as in Figure 6.4. If components 1 and 3 are only partially miscible but component 2 is completely miscible with both components 1 and 3, a diagram such as Figure 6.5 is obtained. x_1^{III} represents the composition of the phase rich in component

6.2 Types of Phase Diagram

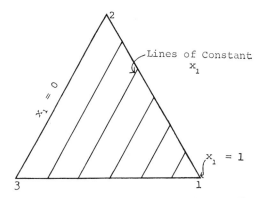

Figure 6.4 Composition representation on triangular diagram.

3 saturated with component 1. Conversely, x_3^I represents the mole fraction of component 3 in saturated solution in component 1. Since the binary pair is only partially miscible, there must be a region in the interior of the triangle with end points at the saturated binary compositions in which two phases exist. If a mixture with overall composition within the two-phase region is prepared, it will separate into two phases lying on a straight line passing through the overall composition point. The line is called a tie-line, shown dotted in Figure 6.5. Its intersections on the boundary of the two-phase region (points A and B) represent the compositions of phases in equilibrium.

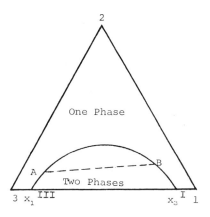

Figure 6.5 Phase diagram with only one partially miscible pair.

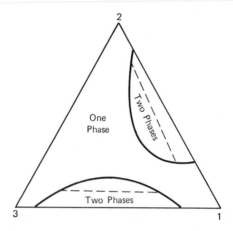

Figure 6.6 Ternary phase diagram with two partially miscible pairs, type 1.

If two of the binaries are partially miscible, a diagram similar to Figure 6.6 is possible, in which there are two regions of two phases. These two regions may be large enough to intersect and merge into a single two-phase region. If the miscibility in both binary systems is roughly of the same magnitude, the diagram may be of the type shown in Figure 6.7. If one region dominates the other, however, Figure 6.8 will be representative of the system.

If all three binary pairs exhibit limited miscibility, there could be three separate two-phase regions on the phase diagram. It is possible that these

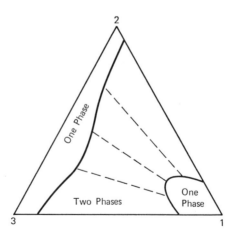

Figure 6.7 Ternary phase diagram with two partially miscible pairs, type 2.

6.2 Types of Phase Diagram

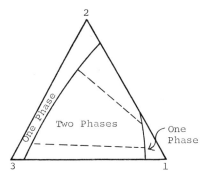

Figure 6.8 Ternary phase diagram with two partially miscible pairs, type 3.

three regions could be sufficiently large to intersect each other, as shown in Figure 6.9. Any mixture having an overall composition within the three-phase region will separate into three distinct layers having compositions represented by points *A*, *B*, and *C* in Figure 6.9.

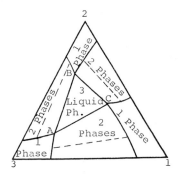

Figure 6.9 Ternary phase diagram exhibiting a three-phase region.

6.2.3 Multicomponent Systems

Systems of more than three components are not readily represented on phase diagrams. The general discussion regarding the appearance of multiphase regions in ternary systems also applies to multicomponent systems, but the number of liquid phases *possible* is equal to the number of components in the system.

6.2.4 Effect of Pressure

In the preceding discussions it has been assumed that the pressure was sufficiently high for no vapor phase to exist. For any given liquid composition there is a unique pressure at which the vapor phase appears. At higher pressures there is no vapor phase. There is some small effect of pressure on the liquid-phase diagrams, but this effect is negligible in most cases. It need be considered only when the pressure is extremely high.

6.2.5 Thermodynamic Interpretation

The basic equations governing liquid–liquid equilibrium are developed in Section 3.5 and are repeated here:

$$(\gamma_i x_i)^{\mathrm{I}} = (\gamma_i x_i)^{\mathrm{II}} = (\gamma_i x_i)^{\mathrm{III}} = \cdots. \tag{6-1}$$

In order for two liquid phases to exist Equation 6-1 must be satisfied simultaneously for *all* components in *all* phases. A trivial solution to these equations is always available:

$$x_i^{\mathrm{I}} = x_i^{\mathrm{II}} = x_i^{\mathrm{III}} = \cdots. \tag{6-2}$$

However, this solution gives all phases as identical in composition and physical state; hence in reality only *one phase* is associated with the solution in Equation 6-2. If two or more liquid phases exist, there must be a solution to Equations 6-1 *other than* Equation 6-2. The maximum number of liquid phases possible with temperature and pressure arbitrarily specified is equal to the number of components present.

The conditions for the formation of two liquid phases in a binary system were discussed in Section 5.2.7 for a system following the symmetrical equations for activity-coefficient dependence on composition. It was shown that two liquid phases cannot exist if $\gamma^\infty \leq 7.39$, whereas partial miscibility exists if $\gamma^\infty > 7.39$. Actually few systems follow the symmetrical equations. Thus, although it is generally true that partial miscibility does not occur if all values of the binary γ^∞ terms are ≤ 7.39, γ^∞ values considerably higher (even approaching 30) can exist in some systems that are completely miscible in all proportions. Unfortunately there is no infallible rule for determining whether two or more liquid phases exist when some γ^∞ values exceed 7.39 except the existence of solutions to Equations 6-1.

In most cases of industrial importance there are at least three components and all equations for γ as a function of composition are highly nonlinear. Consequently there is no simple computational scheme for finding the solutions to Equations 6-1 and the inability to find solutions does not necessarily mean that they do not exist. The situation presents a real dilemma to the

6.2 Types of Phase Diagram

process engineer who wishes to be certain whether a given process will exhibit more than one liquid phase.

Fortunately an alternative criterion to Equations 6-1 permits the certainty desired, but at the expense of increased computational complexity. Equations 6-1 were derived from the more basic criterion (see Chapter 3 for fuller discussion):

$$\mu_i^{\text{I}} = \mu_i^{\text{II}} = \mu_i^{\text{III}} = \cdots . \tag{6-3}$$

It can be shown that Equations 6-3 are equivalent to minimization of the Gibbs free energy of the entire system.

Suppose there is a system with the following characteristics:

N_P phases,
N_i total moles of component i,
N_i^J moles of component i in phase J.

The total Gibbs free energy of such a system of phases and components is given by

$$F = F^{\text{I}} + F^{\text{II}} + \cdots + F^{N_P}.$$

The Gibbs free energy of an individual phase is given by

$$F^J = \sum_i N_i^J \mu_i^J.$$

Thus

$$F = \sum_i (N_i^{\text{I}} \mu_i^{\text{I}} + N_i^{\text{II}} \mu_i^{\text{II}} + \cdots + N_i^{N_P} \mu_i^{N_P}) \tag{6-3}$$

$$= \sum_i \sum_J N_i^J \mu_i^J.$$

The N_i^J are related by

$$N_i = \sum_J N_i^J$$

subject to the restriction $N_i^J \geq 0$. Thus the N_i^J terms can all be considered as independent variables for $J < N_P$, with

$$N_i^{N_P} = N_i - \sum_{J=1}^{N_P - 1} N_i^J. \tag{6-4}$$

A condition of minimum Gibbs free energy implies

$$\frac{\partial F}{\partial N_j^K} = 0 \quad \begin{array}{l}\text{(for all } j \text{ and} \\ \text{for } K < N_P).\end{array} \tag{6-5}$$

By differentiation of Equation 6-3, we obtain

$$\frac{\partial F}{\partial N_j^K} = \mu_j^K + \sum_i N_i^K \frac{\partial \mu_i^K}{\partial N_j^K} + \mu_j^{N_P} \frac{\partial N_j^{N_P}}{\partial N_j^K} + \sum_i N_i^{N_P} \frac{\partial \mu_i^{N_P}}{\partial N_j^K}. \tag{6-6}$$

According to the Gibbs–Duhem equation (Equation 2-16a) for any phase at constant temperature and pressure

$$\sum_i N_i \, d\mu_i = 0; \qquad \sum_i N_i^K \frac{\partial \mu_i^K}{\partial N_j^K} = 0; \qquad \sum_i N_i^{N_P} \frac{\partial \mu_i^{N_P}}{\partial N_j^K} = 0.$$

Hence Equation 6-5 becomes

$$\frac{\partial F}{\partial N_j^K} = \mu_j^K + \mu_j^{N_P} \frac{\partial N_j^{N_P}}{\partial N_j^K}. \tag{6-7}$$

From Equation 6-4 we have

$$\frac{\partial N_j^{N_P}}{\partial N_j^K} = -1. \tag{6-8}$$

By combining Equations 6-8, 6-7, and 6-5, we obtain

$$\mu_j^K - \mu_j^{N_P} = 0,$$

or

$$\mu_j^K = \mu_j^{N_P}. \tag{6-9}$$

In other words, a condition of minimum Gibbs free energy satisfies the criteria for phase equilibrium.

One procedure to determine whether more than one phase exists is to assume a single phase initially ($N_i = N_i^I$). Then assume a second phase initially having all $N_i^{II} = 0$. Make small perturbations in the N_i^I to determine whether any combination gives a lower value of F than $N_i^{II} = 0$. If a second phase is indicated, a search procedure is initiated to find the combination of N_i^I values giving minimum F. When the minimum for two phases is found, initiate a third phase. The procedure is repeated until the introduction of another phase does not reduce F or the number of phases equals the number of components.

Obviously, such a procedure is very tedious and virtually impossible without the aid of a computer.

6.3 EXPERIMENTAL LIQUID–LIQUID EQUILIBRIUM DATA

Liquid–liquid equilibrium data are, undoubtedly, the easiest of all types of phase-equilibrium data to obtain. The appropriate quantities of the components are merely placed together in a container, agitated, and the phases allowed to settle by gravity in a constant-temperature bath. When settling is complete the phases are sampled and analyzed. The only problem arising from the type of phases involved is the necessity of allowing sufficient agitation

6.4 Processing Experimental Data for Liquid–Liquid Equilibria

and contact time to attain equilibrium without formation of a stable emulsion, and sufficient settling time to ensure that only one phase is being sampled at a time. Some problems occur sometimes as a result of the environmental conditions (e.g., sampling at high pressures or low temperatures), but these difficulties are not peculiar to liquid–liquid equilibrium experiments.

6.4 PROCESSING EXPERIMENTAL DATA FOR LIQUID–LIQUID EQUILIBRIA

The reduction of experimental liquid–liquid equilibrium data to a set of mathematical equations presents a unique set of problems. It is not possible to test the data for thermodynamic consistency for two very significant reasons, as follows:

1. The data do not yield values of individual activity coefficients; rather, they give the ratio of two activity coefficients by the equation

$$\frac{\gamma_i^I}{\gamma_i^{II}} = \frac{x_i^{II}}{x_i^I}. \tag{6-10}$$

2. The data do not extend over a continuous composition range; since consistency tests require either integration or differentiation, they cannot be applied to such data. In a binary system, for example, at a given temperature and pressure there are only two distinct compositions available—the mutual solubilities of the two components in each other. All other compositions will either exhibit a single phase only or will separate into the two saturated compositions.

6.4.1 Binary Systems

In the binary case the mutual-solubility data are sufficient to compute the parameters of a two-constant equation by simultaneous solution of the following equations:

$$\gamma_1^I x_1^I = \gamma_1^{II} x_1^{II},$$
$$\gamma_2^I x_2^I = \gamma_2^{II} x_2^{II}.$$

The values of γ_1 and γ_2 are determined by the functional form (ψ) of the equations, the compositions of the solutions represented, and the two parameters A_{12} and A_{21}:

$$\psi(x_1^I, x_2^I, A_{12}, A_{21})x_1^I = \psi(x_1^{II}, x_2^{II}, A_{12}, A_{21})x_1^{II},$$
$$\psi(x_2^I, x_1^I, A_{21}, A_{12})x_2^I = \psi(x_2^{II}, x_1^{II}, A_{21}, A_{12})x_2^{II}. \tag{6-11}$$

Since x_1^I, x_2^I, x_1^{II}, and x_2^{II} are experimentally fixed data, the two simultaneous equations permit the calculation of the two parameters A_{12} and A_{21}. Carlson and Colburn [1] present the solution of Equations 6-11 for the Van Laar equations as

$$\frac{A_{12}}{A_{21}} = \frac{\left(\dfrac{x_1^I}{x_2^I} + \dfrac{x_1^{II}}{x_2^{II}}\right)\left[\dfrac{\ln(x_1^{II}/x_1^I)}{\ln(x_2^I/x_2^{II})}\right] - 2}{\dfrac{x_1^I}{x_2^I} + \dfrac{x_1^{II}}{x_2^{II}} - \dfrac{2x_1^I x_1^{II}}{x_2^I x_2^{II}} \dfrac{\ln(x_1^{II}/x_1^I)}{\ln(x_2^I/x_2^{II})}} \tag{6-12}$$

$$A_{12} = \frac{\ln(x_1^{II}/x_1^I)}{[1 + (A_{12}/A_{21})(x_1^I/x_2^I)]^{-2} - [1 + (A_{12}/A_{21})(x_1^{II}/x_2^{II})]^{-2}}. \tag{6-13}$$

Thus one could use Equation 6-12 with mutual-solubility data to obtain the ratio (A_{12}/A_{21}). Then by applying Equation 6-13 A_{12} is known, and A_{21} is calculated by

$$A_{21} = \frac{A_{12}}{(A_{12}/A_{21})}.$$

Similar expressions are available for the Margules equations:

$$(A_{21} - A_{12}) = \frac{\ln\left(\dfrac{x_2^{II} x_1^I}{x_1^{II} x_2^I}\right)}{1 + \dfrac{2[x_1^I x_2^I(x_1^{II} - x_2^{II}) + x_1^{II} x_2^{II}(x_1^I - x_2^I)]}{(x_1^I + x_1^{II})(x_2^I + x_2^{II})}} \tag{6-14}$$

$$A_{12} = \ln\frac{x_1^{II}}{x_1^I} - 2(A_{21} - A_{12})\left[\frac{(x_2^I)^2 x_1^I - (x_2^{II})^2 x_1^{II}}{(x_2^I)^2 - (x_2^{II})^2}\right]. \tag{6-15}$$

The Wilson equations cannot be used for liquid–liquid equilibria since they always predict complete miscibility. Three-constant equations have too many constants to allow the retrieval of the binary parameters from mutual-solubility data.

The parameters obtained from mutual-solubility data allow for no statistical treatment since only one set of data is used, and the determination of the parameters is by exact solution of the equations for that set of data. It is therefore not surprising that the parameters obtained from mutual-solubility data are often quite different from those obtained by infinite-dilution activity-coefficient measurements or from vapor–liquid equilibrium data, unless the solubilities are extremely small (less than 1 mole %).

The differences between the results one obtains by the mutual-solubility method are, however, more severe than can be explained on the basis of statistical error in either the mutual-solubility data or the corresponding vapor–liquid equilibrium data. The discrepancies actually indicate that, for

6.4 Processing Experimental Data for Liquid–Liquid Equilibria

systems that exhibit limited liquid-phase miscibility, the two-constant isothermal equations do not adequately represent the liquid-phase activity coefficients. Three-constant equations are required, and constants for these cannot be obtained from mutual-solubility data. They can, however, be obtained from vapor–liquid equilibrium data by the methods presented in the preceding chapter if such data are available. If some measurement or estimate of one or both of the activity coefficients at infinite dilution is available, as well as mutual-solubility data, the third constant can be estimated in a three-constant equation. The exact type of the calculation depends on the available data in each specific case; thus all possible combinations cannot be covered here, but two examples are included to illustrate the procedure.

Example 6-1

Develop a procedure for determining the three isothermal constants of Equation 4-74 when one infinite-dilution activity coefficient is known and mutual-solubility data are available.

Solution. The equations to be used are   *Van Laar:*

$$\ln \gamma_1 = A_{12} Z_2^2 \left[1 + \frac{D_{12}}{A_{12}} (3Z_1^2 - 2Z_1) \right]$$

and

$$\ln \gamma_2 = A_{21} Z_1^2 \left[1 + \frac{D_{21}}{A_{21}} (3Z_2^2 - 2Z_2) \right],$$

where

$$\frac{D_{12}}{D_{21}} = \frac{A_{12}}{A_{21}},$$

$$Z_1 = \frac{A_{12} x_1}{A_{12} x_1 + A_{21} x_2},$$

$$Z_2 = 1 - Z_1.$$

If γ_1^∞ is known, A_{12} can be determined, since at $x_1 = 0$, $Z_1 = 0$ and $Z_2 = 1$, and therefore $A_{12} = \ln \gamma_1^\infty$.

The other two parameters must be determined from the equations

$$\gamma_1^{\text{I}} x_1^{\text{I}} = \gamma_1^{\text{II}} x_1^{\text{II}}; \qquad \gamma_2^{\text{I}} x_2^{\text{I}} = \gamma_2^{\text{II}} x_2^{\text{II}},$$

where x_1^{I}, x_1^{II}, x_2^{I}, and x_2^{II} represent the mutual-solubility data (mole fractions of the components in phases I and II, respectively). By letting

$$C_1 = C_2 = \frac{D_{12}}{A_{12}} = \frac{D_{21}}{A_{21}},$$

the activity-coefficient equations become

$$\ln \gamma_1 = A_{12} Z_2^2 [1 + C_1(3Z_1^2 - 2Z_1)],$$
$$\ln \gamma_2 = A_{21} Z_1^2 [1 + C_2(3Z_2^2 - 2Z_2)].$$

The phase-equilibrium equations can be written

$$\ln \gamma_1^{\mathrm{I}} x_1^{\mathrm{I}} = \ln \gamma_1^{\mathrm{II}} x_1^{\mathrm{II}},$$
$$\ln \gamma_2^{\mathrm{I}} x_2^{\mathrm{I}} = \ln \gamma_2^{\mathrm{II}} x_2^{\mathrm{II}}.$$

The phase-equilibrium equation for component 1, combined with the activity-coefficient equation, gives

$$A_{12}(Z_2^{\mathrm{I}})^2 \{1 + C_1[3(Z_1^{\mathrm{I}})^2 - 2Z_1^{\mathrm{I}}]\} + \ln x_1^{\mathrm{I}}$$
$$= A_{12}(Z_2^{\mathrm{II}})^2 \{1 + C_1[3(Z_1^{\mathrm{II}})^2 - 2Z_1^{\mathrm{II}}]\} + \ln x_2^{\mathrm{II}},$$

which can be solved for C_1:

$$C_1 = \frac{[(Z_2^{\mathrm{II}})^2 - (Z_2^{\mathrm{I}})^2] + \dfrac{\ln(x_1^{\mathrm{II}}/x_1^{\mathrm{I}})}{A_{12}}}{[3(Z_1^{\mathrm{I}})^2 - 2Z_1^{\mathrm{I}}](Z_2^{\mathrm{I}})^2 - [3(Z_1^{\mathrm{II}})^2 - 2Z_1^{\mathrm{II}}](Z_2^{\mathrm{II}})^2}.$$

Similarly, based on component 2,

$$C_2 = \frac{[(Z_1^{\mathrm{II}})^2 - (Z_1^{\mathrm{I}})^2] + \dfrac{\ln(x_2^{\mathrm{II}}/x_2^{\mathrm{I}})}{A_{21}}}{[3(Z_2^{\mathrm{I}})^2 - 2Z_2^{\mathrm{I}}](Z_1^{\mathrm{I}})^2 - [3(Z_2^{\mathrm{II}})^2 - 2Z_2^{\mathrm{II}}](Z_1^{\mathrm{II}})^2}$$

Since $C_1 = C_2$, one can assume various values of A_{21} and compute $C_1 - C_2$. Then $C_1 - C_2$ is plotted against A_{21}. The value of A_{21} for which $C_1 - C_2 = 0$ gives the correct value of A_{21}, and the corresponding value of C is the correct value for the third constant.

Example 6-2

Develop a procedure for determining the parameters of Equation 4-74 when both infinite-dilution activity coefficients are known and the solubility of component 1 in component 2 is known.

Solution. In this example the values of A_{12} and A_{21} are given (analogy to Example 6-1) by

$$A_{12} = \ln \gamma_1^\infty; \qquad A_{21} = \ln \gamma_2^\infty.$$

The equations for C_1 and C_2 are valid, as in Example 6-1. However, the data here give only x_1^{I} and x_2^{I}. The term x_1^{II} is the unknown quantity ($x_2^{\mathrm{II}} = 1 - x_1^{\mathrm{II}}$). Therefore values of x_1^{II} are assumed, $C_1 - C_2$ is plotted against x_1^{II}, and the value of x_1^{II} for which $C_1 - C_2 = 0$ is read from the plot.

6.4 Processing Experimental Data for Liquid–Liquid Equilibria

Of course, the plotting procedure could be replaced by a numerical root-finding technique and a computer program.

Procedures similar to those developed in Examples 6-1 and 6-2 can be applied to any of the three-constant equations developed in Chapter 4 and frequently γ^∞ values can be estimated by one of the correlations presented in Chapter 4 or measured by one of the methods discussed in Chapter 5. The procedure must, of course, be adapted to the type of data available.

Since the data-analysis procedures are all numerical, iterative methods, Table 6.1 can be quite useful for obtaining graphical solutions. Each table represents a specific value of the four-suffix correction factor C with the values of A_{12} and A_{21} corresponding to various mutual-solubility values tabulated. Thus, for example, if C is 0.5 and the mutual solubilities are 0.05 for component 1 in component 2 and 0.1 for component 2 in component 1, we find in the table labeled $C = 0.5$, in columns $x_1{}^I = 0.05$ and $x_1{}^{II} = 0.9$, that $A_{12} = 3.364$ and $A_{21} = 2.761$. By means of judicious cross-plotting, the tables can be adapted to a variety of types of binary-data-analysis problems.

Temperature variation is important in liquid–liquid equilibrium. Since solubilities vary with temperature, it is evident that the activity-coefficient-equation parameters vary with temperature. There is no completely satisfactory theoretical form for the variation of parameters, but the following empirical equations are usually satisfactory. For equations based on the Wohl formulation

$$A = a + \frac{b}{T} + cT, \tag{6-16}$$

where T is the absolute temperature. In most cases c can be assumed to be 0. For equations based on the Wilson formulation

$$\ln A = a + \frac{b}{T}, \tag{6-17}$$

where A represents any of the parameters of the activity-coefficient equation.

It has been suggested that when mutual-solubility data are available over a range of temperature, a regression analysis of the mutual-solubility data be run and all the constants (a, b, c) of Equations 6-16 or 6-17 be determined by fitting calculated mutual solubilities against experimental values. This can be done, but the author's experience has been that much better results are obtained if the parameters are determined at specific temperatures and then fitted for temperature coefficients. Particularly disastrous results can be obtained if parameters obtained by the mutual-solubility regression analysis are used in vapor–liquid equilibrium calculations.

Table 6.1 Van Laar Constants from Mutual Solubilities

C = 0

x_1^I	x_1^{II}	A_{12}	A_{21}
.01	.05	3.537	.1677
	.10	3.747	.2704
	.20	4.017	.4585
	.35	4.265	.7390
	.50	4.430	1.047
	.65	4.549	1.421
	.80	4.639	1.956
	.90	4.682	2.583
	.95	4.695	3.205
	.99	4.689	4.689
.05	.10	3.190	.4438
	.20	3.223	.6599
	.35	3.275	.9545
	.50	3.310	1.258
	.65	3.328	1.612
	.80	3.325	2.108
	.90	3.302	2.689
	.95	3.272	3.272
	.99	3.205	4.695
.10	.20	2.979	.8135
	.35	2.934	1.107
	.50	2.902	1.399
	.65	2.865	1.734
	.80	2.810	2.199
	.90	2.747	2.747
	.95	2.689	3.302
	.99	2.583	4.682
.20	.35	2.628	1.321
	.50	2.523	1.590
	.65	2.425	1.891
	.80	2.310	2.310
	.90	2.199	2.810
	.95	2.108	3.325
	.99	1.956	4.639
	.50	2.218	1.804
	.65	2.063	2.063
	.80	1.891	2.425
	.90	1.734	2.865
	.95	1.612	3.328
	.99	1.421	4.549
.50	.65	1.804	2.218
	.80	1.590	2.523
	.90	1.399	2.902
	.95	1.258	3.310
	.99	1.047	4.430
	.80	1.321	2.628
	.90	1.107	2.934
	.95	.9545	3.275
	.99	.7390	4.265
.80	.90	.8135	2.979
	.95	.6599	3.223
	.99	.4585	4.017
.90	.95	.4438	3.190
	.99	.2704	3.747
	.99	.1677	3.537

Table 6.1 (*Continued*)

$c = 0.5$

x_1^I	x_1^{II}	A_{12}	A_{21}
.01	.05	3.694	.1970
	.10	3.883	.3113
	.20	4.126	.5145
	.35	4.349	.8082
	.50	4.494	1.122
	.65	4.599	1.496
	.80	4.675	2.021
	.90	4.710	2.634
	.95	4.718	3.242
	.99	4.704	4.704
.05	.10	3.355	.5235
	.20	3.372	.7624
	.35	3.402	1.073
	.50	3.418	1.379
	.65	3.419	1.726
	.80	3.401	2.204
	.90	3.364	2.761
	.95	3.324	3.324
	.99	3.242	4.718
.10	.20	3.138	.9441
	.35	3.078	1.253
	.50	3.032	1.546
	.65	2.981	1.870
	.80	2.910	2.311
	.90	2.831	2.831
	.95	2.761	3.364
	.99	2.634	4.710
.20	.35	2.788	1.495
	.50	2.676	1.761
	.65	2.570	2.048
	.80	2.441	2.441
	.90	2.311	2.910
	.95	2.204	3.401
	.99	2.021	4.675
	.50	2.390	1.987
	.65	2.233	2.233
	.80	2.048	2.570
	.90	1.869	2.981
	.95	1.726	3.419
	.99	1.496	4.599
.50	.65	1.987	2.390
	.80	1.761	2.677
	.90	1.546	3.032
	.95	1.379	3.418
	.99	1.122	4.494
	.80	1.495	2.788
	.90	1.253	3.078
	.95	1.073	3.402
	.99	.8082	4.349
.80	.90	.9434	3.138
	.95	.7622	3.372
	.99	.5145	4.126
.90	.95	.5225	3.356
	.99	.3111	3.883
	.99	.1967	3.694

Table 6.1 (*Continued*)

C = 1.0

x_1^I	x_1^{II}	A_{12}	A_{21}
.01	.05	3.830	.2389
	.10	4.007	.3692
	.20	4.231	.5918
	.35	4.430	.8996
	.50	4.557	1.217
	.65	4.647	1.585
	.80	4.710	2.095
	.90	4.737	2.689
	.95	4.740	3.281
	.99	4.720	4.720
.05	.10	3.491	.6401
	.20	3.498	.9123
	.35	3.514	1.240
	.50	3.518	1.542
	.65	3.507	1.870
	.80	3.475	2.314
	.90	3.427	2.839
	.95	3.379	3.379
	.99	3.281	4.740
.10	.20	3.268	1.139
	.35	3.202	1.467
	.50	3.150	1.749
	.65	3.093	2.043
	.80	3.013	2.441
	.90	2.922	2.922
	.95	2.839	3.427
	.99	2.689	4.737
.20	.35	2.925	1.753
	.50	2.821	2.000
	.65	2.719	2.247
	.80	2.586	2.586
	.90	2.441	3.013
	.95	2.314	3.475
	.99	2.095	4.710
	.50	2.571	2.230
	.65	2.432	2.432
	.80	2.247	2.719
	.90	2.043	3.093
	.95	1.870	3.507
	.99	1.585	4.647
.50	.65	2.230	2.572
	.80	2.000	2.821
	.90	1.749	3.150
	.95	1.542	3.518
	.99	1.217	4.557
	.80	1.752	2.926
	.90	1.467	3.202
	.95	1.240	3.514
	.99	.8995	4.430
.80	.90	1.138	3.269
	.95	.9120	3.498
	.99	.5917	4.231
.90	.95	.6389	3.491
	.99	.3691	4.007
	.99	.2386	3.830

Table 6.1 (*Continued*)

$c = 3$

x_1^I	x_1^{II}	A_{12}	A_{21}
.01	.05	3.970	.3529
	.10	4.174	.5455
	.20	4.407	.8500
	.35	4.583	1.206
	.50	4.678	1.512
	.65	4.738	1.834
	.80	4.777	2.277
	.90	4.789	2.811
	.95	4.784	3.363
	.99	4.752	4.752
.05	.10	3.593	.9502
	.20	3.618	1.368
	.35	3.648	1.827
	.50	3.653	2.130
	.65	3.644	2.331
	.80	3.614	2.604
	.90	3.556	3.016
	.95	3.493	3.493
	.99	3.363	4.784
.10	.20	3.358	1.720
	.35	3.303	2.218
	.50	3.268	2.530
	.65	3.248	2.649
	.80	3.214	2.785
	.90	3.121	3.121
	.95	3.016	3.556
	.99	2.811	4.789
.20	.35	2.986	2.707
	.50	2.905	3.006
	.65	2.910	3.002
	.80	2.937	2.937
	.90	2.785	3.214
	.95	2.604	3.614
	.99	2.277	4.777
	.50	2.647	3.397
	.65	2.962	2.962
	.80	3.002	2.910
	.90	2.649	3.248
	.95	2.331	3.644
	.99	1.834	4.738
.50	.65	3.398	2.646
	.80	3.006	2.905
	.90	2.530	3.268
	.95	2.130	3.653
	.99	1.512	4.678
	.80	2.707	2.986
	.90	2.218	3.303
	.95	1.827	3.648
	.99	1.206	4.583
.80	.90	1.720	3.358
	.95	1.367	3.618
	.99	.8499	4.407
.90	.95	.9491	3.593
	.99	.5454	4.174
	.99	.3526	3.970

Table 6.1 (*Continued*)

$c = 5$

x_1^I	x_1^{II}	A_{12}	A_{21}
.01	.05	3.918	.4537
	.10	4.176	.7345
	.20	4.482	1.219
	.35	4.712	1.800
	.50	4.799	2.147
	.65	4.822	2.291
	.80	4.837	2.532
	.90	4.839	2.956
	.95	4.827	3.453
	.99	4.784	4.784
.05	.10	3.505	1.200
	.20	3.563	1.805
	.35	3.639	2.596
	.50	3.676	3.238
	.65	3.681	3.451
	.80	3.708	3.099
	.90	3.684	3.231
	.95	3.615	3.615
	.99	3.453	4.827
.10	.20	3.270	2.228
	.35	3.236	3.055
	.50	3.215	3.781
	.65	3.196	4.193
	.80	3.318	3.464
	.90	3.349	3.349
	.95	3.231	3.684
	.99	2.956	4.839
.20	.35	2.872	3.676
	.50	2.768	4.413
	.65	2.696	4.986
	.80	3.398	3.398
	.90	3.464	3.318
	.95	3.099	3.708
	.99	2.532	4.837
	.50	2.406	5.073
	.65	3.786	3.786
	.80	4.986	2.696
	.90	4.193	3.196
	.95	3.451	3.681
	.99	2.291	4.822
.50	.65	5.075	2.406
	.80	4.413	2.768
	.90	3.781	3.215
	.95	3.238	3.676
	.99	2.147	4.799
	.80	3.677	2.872
	.90	3.055	3.236
	.95	2.596	3.639
	.99	1.800	4.712
.80	.90	2.227	3.270
	.95	1.805	3.563
	.99	1.219	4.482
.90	.95	1.199	3.505
	.99	.7344	4.176
	.99	.4535	3.918

Table 6.1 (*Continued*)

$c = 10$

x_1^{I}	x_1^{II}	A_{12}	A_{21}
.01	.05	3.555	.5496
	.10	3.861	.9472
	.20	4.258	1.753
	.35	4.633	3.083
	.50	4.881	4.619
	.65	5.031	6.167
	.80	4.933	3.969
	.90	4.924	3.363
	.95	4.910	3.661
	.99	4.849	4.849
.05	.10	3.136	1.413
	.20	3.228	2.235
	.35	3.362	3.537
	.50	3.470	5.083
	.65	3.553	7.063
	.80	3.595	9.106
	.90	3.871	3.962
	.95	3.887	3.887
	.99	3.661	4.910
.10	.20	2.923	2.677
	.35	2.924	3.920
	.50	2.943	5.378
	.65	2.957	7.298
	.80	2.956	9.957
	.90	3.924	3.924
	.95	3.962	3.870
	.99	3.363	4.924
.20	.35	2.546	4.560
	.50	2.459	5.852
	.65	2.388	7.575
	.80	4.951	4.951
	.90	9.957	2.956
	.95	9.107	3.595
	.99	3.968	4.933
	.50	2.084	6.532
	.65	8.538	8.538
	.80	7.575	2.388
	.90	7.298	2.957
	.95	7.063	3.553
	.99	6.167	5.031
.50	.65	6.534	2.083
	.80	5.853	2.459
	.90	5.378	2.943
	.95	5.083	3.470
	.99	4.619	4.881
	.80	4.561	2.546
	.90	3.920	2.924
	.95	3.537	3.362
	.99	3.083	4.633
.80	.90	2.677	2.923
	.95	2.235	3.228
	.99	1.753	4.258
.90	.95	1.412	3.137
	.99	.9472	3.861
	.99	.5494	3.555

6.4.2 Multicomponent Systems

In attempting to fit multicomponent liquid–liquid equilibrium data we encounter directly the problems of insufficient data and limited choice of equations for activity coefficients. Insufficiency of the data can be alleviated by supplementary vapor–liquid equilibrium data; indeed, it is advisable to obtain as many of the parameters as possible from vapor–liquid equilibrium data, with only the remaining ones obtained by regression analysis on the liquid–liquid data.

The choice of an equation for representing activity coefficients is, however, a formidable problem. Systems forming multiple liquid phases have, inherently, very large positive deviations from ideal-solution behavior. In the analysis reported in Section 5.4.1 only the Wilson and Van Laar equations give satisfactory results for binary systems with large positive deviations. The Wilson equations cannot be used because they cannot, mathematically, predict liquid-phase separation. The Van Laar equations are, as pointed out in Section 4.3.3, restricted by the requirement

$$\frac{A_{ij}}{A_{ji}} = \frac{A_{ik}}{A_{ki}} \frac{A_{kj}}{A_{jk}}. \tag{6-18}$$

This requirement is not met by very many systems.

The three-constant equations are a possibility, but those not based on the Van Laar assumptions (hence subject to the same restrictions) usually require more than three constants to represent the binary systems adequately. The inclusion of higher order terms introduces the mathematical possibility of multiple solutions to the phase-equilibrium equations (Equations 6-1). Hence we appear to be on the horns of a dilemma. Black [2] and R. H. Fariss† have formulated their equations to be independent of the restrictions of Equation 6-18. Black's equations, however, contain higher order terms with the possibility of multiple solutions. At the time of this writing the multicomponent version of the Fariss equations has not been presented.

A partial solution to this dilemma has been successfully applied to process calculations within the Monsanto Company over a period of years. Each component is assumed to behave as though it were present in a pseudobinary system obeying the Van Laar equations:

$$\ln \gamma_i = A_i(1 - Z_i)^2[1 + C_i(Z_i^2 - \tfrac{2}{3}Z_i)]. \tag{6-19}$$

The parameters A_i, B_i, and C_i are determined from binary-system parameters as follows:

---

†Private communication to the author, 1963.

6.5 Process Calculations

$$A_i = \frac{\sum_j A_{ij} x_j}{1 - x_i},$$

$$B_i = \frac{\sum_j A_{ji} x_j}{1 - x_i},$$

$$C_i = \frac{\sum_j C_{ij} x_j}{1 - x_i},$$

$$C_{ij} = C_{ji} = 3\frac{D_{ij}}{A_{ij}} = 3\frac{D_{ji}}{A_{ji}}.$$

(6-20)

The above equations† have the advantage of avoiding the restrictions of Equation 6-18, reducing to the Van Laar equations for all of the component binaries, requiring no specific ternary parameters, and being simpler in form than the multicomponent Wohl formulations of Chapter 4. Their prime disadvantage is that they do not rigorously obey the Gibbs–Duhem equation, although data calculated by using them are usually reasonably consistent.

The procedure recommended by the author for fitting liquid–liquid equilibrium data, then, is as follows:

1. Obtain as many parameters as possible from vapor–liquid equilibria or correlations of γ^∞.
2. Obtain the remaining parameters by nonlinear regression analysis (Section 5.4.1), using distribution coefficient (x^I/x^{II}) or calculated solubility as the objective function.

6.5 PROCESS CALCULATIONS

The basic question to be answered in process calculations of liquid–liquid equilibrium is as follows:

Given the temperature, pressure, and composition of a liquid mixture, how many liquid phases will form and what is the composition of each?

Although, theoretically, it is possible for as many liquid phases to exist as there are components present, most commercially significant situations

---

†There is some experience to show that a better averaging method can be obtained by using $A_i = \sum_j A_{ij} x_j^N/(1 - x_i)^N$ etc., where N has a value of about 1.5. For the present, the evidence is not conclusive and Equations (6-20) are used here.

involve only one or two liquid phases. Industrial equipment for handling more than two liquid phases becomes quite complex; therefore, if a third liquid phase is possible, it is quite important that we know about it.

6.5.1 Binary Systems

If more than two components are involved, liquid–liquid equilibrium computations are usually sufficiently complex to require the use of a computer. If programs are available for multicomponent calculations, they can, of course, be applied to binary systems as well. However, a rather simple graphical procedure can be used to calculate mutual solubility in binary systems, given the appropriate parameters for the activity-coefficient equations.

In the binary system the equations to be solved simultaneously are

and
$$(\gamma_1 x_1)^{\mathrm{I}} = (\gamma_1 x_1)^{\mathrm{II}}$$
$$(\gamma_2 x_2)^{\mathrm{I}} = (\gamma_2 x_2)^{\mathrm{II}},$$
(6-21)

where
$$x_1^{\mathrm{I}} + x_2^{\mathrm{I}} = 1 \quad \text{and} \quad x_1^{\mathrm{II}} + x_2^{\mathrm{II}} = 1,$$

subject to the conditions
$$0 \le x_1^{\mathrm{I}} \le 1 \quad \text{and} \quad 0 \le x_1^{\mathrm{II}} \le 1.$$

If $\gamma_1 x_1$ and $\gamma_2 x_2$ are calculated as functions of x_1, a plot of $\gamma_1 x_1$ against $\gamma_2 x_2$ will loop and cross itself if a solution exists. The two values of x_1 at the intersection of the two branches of the curve represent x_1^{I} and x_1^{II}. The following example illustrates this method:

Example 6-3

Aniline and methylcyclohexane, in binary solution, have infinite-dilution activity coefficients of 16.1 and 11.14, respectively, at 25°C. Using a three-suffix Van Laar equation, determine their mutual solubility.

Solution. The constants for the Van Laar equations (Equation 4-58) are given by
$$A_{12} = \ln \gamma_1^\infty = \ln(16.1) = 2.78,$$
$$A_{21} = \ln \gamma_2^\infty = \ln(11.14) = 2.41.$$

Equations 4-62 give, for γ,
$$\gamma_1 = \exp(A_{12} Z_2^2); \quad \gamma_2 = \exp(A_{21} Z_1^2),$$
where
$$Z_1 = \frac{A_{12} x_1}{A_{12} x_1 + A_{21} x_2}; \quad Z_2 = 1 - Z_1.$$

6.5 Process Calculations

For $x_1 = 0.01$ the calculations give

$$Z_1 = \frac{2.78 \times 0.01}{(2.78 \times 0.01) + (2.41 \times 0.99)} = 0.0115,$$

$$Z_2 = 1 - 0.0115 = 0.9885;$$
$$\gamma_1 = \exp(2.78 \times 0.9885^2) = 15.12,$$
$$\gamma_2 = \exp(2.41 \times 0.0115^2) = 1.0003;$$
$$\gamma_1 x_1 = 15.12 \times 0.01 = 0.1512,$$
$$\gamma_2 x_2 = 1.0003 \times 0.99 = 0.9903.$$

Values calculated for the complete range of x_1 are given in Table 6.2. The two plots convenient for the solution are shown in Figure 6.10. Plot *a* gives $\gamma_1 x_1$ against $\gamma_2 x_2$, with some corresponding values of x_1 shown on the curve. The values of $x_1{}^I$ and $x_1{}^{II}$ are more conveniently read from plot *b* ($\gamma_1 x_1$ against x_1). The value of $\gamma_1 x_1$ at the intersection of the loop is found on plot *b*. There are three such values, but only two correspond to the loop intersection. Thus the calculated solubilities are

$$x_1{}^I \text{ (aniline in methylcyclohexane)} = 0.105,$$
$$x_2{}^{II} \text{ (methylcyclohexane in aniline)} = 1 - x_1{}^{II} = 1 - 0.848 = 0.152.$$

An alternative solution to Example 6-3 could be obtained by making use of Table 6.1.

Alternative Solution. Table 6.1 for $C = 0$ is shown plotted in Figure 6.11. Values of A_{12} and A_{21} are plotted against $x_1{}^{II}$ at constant values of $x_1{}^I$. From the preceding solution A_{21} has a value of 2.41. A horizontal line is drawn for a value $A_{21} = 2.41$. This line intersects each of the A_{21} curves, from which the corresponding value of $x_1{}^{II}$ can be read. Then the A_{12} curve for the same value of $x_1{}^I$ and, corresponding to the $x_1{}^{II}$ value, can be read. Thus a table of A_{12}, $x_1{}^I$, and $x_1{}^{II}$ values can be constructed corresponding to a value of $A_{21} = 2.41$:

$x_1{}^I$	$x_1{}^{II}$	A_{12}
0.01	0.879	4.68
0.05	0.863	3.31
0.1	0.85	2.8
0.2	0.827	2.29

Now the values of $x_1{}^I$ and $x_1{}^{II}$ are plotted against A_{12}. The values corresponding to $A_{12} = 2.78$ are the solution. From Figure 6.12 $x_1{}^I = 0.105$ and $x_1{}^{II} = 0.848$.

Table 6.2 Calculated Data for Solution of Example 6-3
(Aniline–Methylcyclohexane at 25°C)

x_1	γ_1	γ_2	$\gamma_1 x_1$	$\gamma_2 x_2$
.01	15.1247	1.0003	.1512	.9903
.02	14.205	1.0013	.2841	.9813
.03	13.3534	1.0029	.4006	.9728
.04	12.5644	1.0051	.5026	.9649
.05	11.8327	1.0079	.5916	.9575
.06	11.1537	1.0114	.6692	.9507
.07	10.523	1.0155	.7366	.9444
.08	9.9368	1.0202	.7949	.9386
.09	9.3914	1.0256	.8452	.9333
.10	8.8837	1.0316	.8884	.9284
.11	8.4107	1.0382	.9252	.924
.12	7.9697	1.0455	.9564	.9201
.13	7.5582	1.0535	.9826	.9165
.14	7.1741	1.0621	1.0044	.9134
.15	6.8151	1.0714	1.0223	.9107
.16	6.4795	1.0813	1.0367	.9083
.17	6.1654	1.092	1.0481	.9064
.18	5.8714	1.1034	1.0568	.9048
.19	5.5959	1.1155	1.0632	.9035
.20	5.3376	1.1283	1.0675	.9027
.25	4.2643	1.2043	1.0661	.9032
.30	3.4725	1.3018	1.0418	.9113
.35	2.8801	1.4245	1.008	.9259
.40	2.4311	1.5769	.9724	.9461
.45	2.0871	1.765	.9392	.9708
.50	1.8211	1.9966	.9105	.9983
.55	1.614	2.2816	.8877	1.0267
.60	1.452	2.6325	.8712	1.053
.65	1.3252	3.0653	.8614	1.0729
.70	1.2263	3.6008	.8584	1.0803
.75	1.15	4.2655	.8625	1.0664
.80	1.0922	5.0932	.8738	1.0186
.85	1.0502	6.1281	.8926	.9192
.86	1.0434	6.3647	.8974	.8911
.87	1.0372	6.6123	.9024	.8596
.88	1.0316	6.8715	.9078	.8246
.89	1.0264	7.1428	.9135	.7857
.90	1.0217	7.427	.9195	.7427
.91	1.0175	7.7246	.9259	.6952
.92	1.0138	8.0363	.9327	.6429
.93	1.0105	8.3628	.9398	.5854
.94	1.0077	8.705	.9472	.5223
.95	1.0053	9.0636	.955	.4532
.96	1.0034	9.4394	.9632	.3776
.97	1.0019	9.8334	.9718	.295
.98	1.0008	10.2465	.9808	.2049
.99	1.0002	10.6797	.9902	.1068

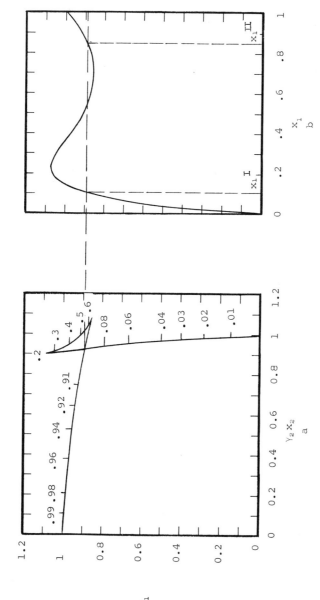

Figure 6.10 Solution of Example 6-3. Mutual solubility of aniline-methylcyclohexane at 25°C.

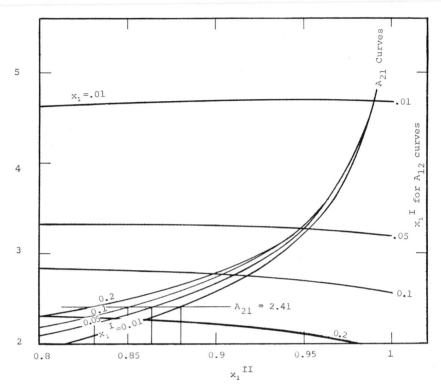

Figure 6.11 Alternative solution to Example 6-3.

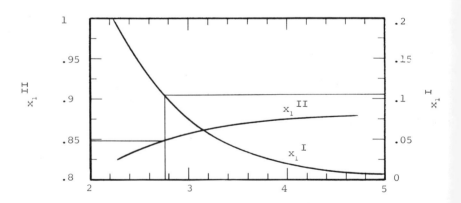

Figure 6.12 Mutual-solubility computation, Example 6-3.

6.5.2 Ternary Systems

Systems of three components can exist in one, two, or three liquid phases. The equations to be solved for two liquid phases are

$$(\gamma_i x_i)^\mathrm{I} = (\gamma_i x_i)^\mathrm{II} \quad (i = 1, 2, 3),$$

$$\sum_{i=1}^{3} x_i^\mathrm{I} = 1, \quad (6\text{-}22)$$

$$\sum_{i=1}^{3} x_i^\mathrm{II} = 1.$$

The equations for three liquid phases are quite similar:

$$(\gamma_i x_i)^\mathrm{I} = (\gamma_i x_i)^\mathrm{II} = (\gamma_i x_i)^\mathrm{III} \quad (i = 1, 2, 3),$$

$$\sum_{i=1}^{3} x_i^\mathrm{I} = 1,$$

$$\sum_{i=1}^{3} x_i^\mathrm{II} = 1, \quad (6\text{-}23)$$

$$\sum_{i=1}^{3} x_i^\mathrm{III} = 1.$$

The absence of a solution to either Equation 6-22 or 6-23 indicates a single liquid phase.

Rigorous solutions to the equations, regardless of the form of equations used for activity coefficients, usually require the aid of modern computers; however, if the assumption can be made that two of the components are completely immiscible, calculations of sufficient accuracy can often be made by hand. The rigorous and approximate solutions are both discussed, with Equations 6-19 being used to calculate activity coefficients.

Two Liquid Phases. The rigorous solution to Equations 6-22 is usually done by assuming that the feed of composition x_i^F splits into two liquid phases of composition x_i^I and x_i^II, with the fraction ϕ of the total feed (mole basis) going into phase I. Thus a material balance gives

$$x_i^F = \phi x_i^\mathrm{I} + (1 - \phi) x_i^\mathrm{II}. \quad (6\text{-}24)$$

Equation 6-22 can be rearranged to give

$$x_i^\mathrm{I} = \frac{\gamma_i^\mathrm{II}}{\gamma_i^\mathrm{I}} x_i^\mathrm{II} = K_i x_i^\mathrm{II} \quad (6\text{-}25)$$

where

$$K_i \equiv \frac{\gamma_i^\mathrm{II}}{\gamma_i^\mathrm{I}}.$$

Equation 6-25, applied to Equation 6-24, yields

$$x_i^F = K_i \phi x_i^{II} + (1 - \phi)x_i^{II},$$

or

$$x_i^{II} = \frac{x_i^F}{1 + (K_i - 1)\phi}. \tag{6-26}$$

Thus, if two liquid phases exist, we must find a set of solutions to Equation 6-26 for which

$$\sum x_i^{II} = 1. \tag{6-27}$$

If Equation 6-27 holds, the material balance of Equation 6-24 will ensure that $\sum x_i^I = 1$.

To solve the equations one normally assumes an initial set of values for x_i^I and an initial value of ϕ. Then x_i^{II} values are calculated by Equation 6-24. The γ_i^I and γ_i^{II} values are calculated from the compositions thus obtained. New values of x_i^{II} and x_i^I are then calculated from Equations 6-26 and 6-25, respectively. If Equation 6-27 is not satisfied, ϕ is adjusted and Equations 6-26 and 6-25 are used again for x_i^{II} and x_i^I, respectively, until Equation 6-27 is satisfied.

Some modification of the Newton–Raphson technique is usually used to modify the ϕ value after each iteration. If we let

$$\psi = \sum x_i^{II} - 1, \tag{6-28}$$

then

$$\frac{d\psi}{d\phi} = \sum \frac{dx_i^{II}}{d\phi}.$$

The values of $dx_i^{II}/d\phi$ are obtained by differentiating Equation 6-26, neglecting the dependence of K_i on composition:

$$\frac{dx_i^{II}}{d\phi} \simeq -\frac{(K_i - 1)x_i^F}{[1 + (K_i - 1)\phi]^2}. \tag{6-29}$$

Since we wish to make $\psi = 0$, the Newton–Raphson technique gives successive estimates of ϕ by

$$\phi_{\text{new}} = \phi_{\text{old}} - \frac{\psi}{d\psi/d\phi}. \tag{6-30}$$

This procedure is illustrated in the following example:

Example 6-4

Compute and construct the ternary liquid–liquid phase diagram for the system acetone–methyl isobutyl ketone–water at 25°C. Acetone and methyl

```
110 DIMENSIØN E(10),R(10),F(10),A(10,10)
120 DIMENSIØN C(10,10),GE(10),GR(10),EK(10)
130 DIMENSIØN WM(10)
140 STRT1: INPUT,NC,F(1),E(1); READ,(F(I),E(I),I=2,NC)
160 READ,(WM(I),I=1,NC)
170 DØ THIS, I=1,NC
180 THIS: R(I)=F(I)-E(I)
190 SUMF = 0.
200 DØ FSUM, I=1,NC
210 FSUM: SUMF = SUMF + F(I)
220 DØ XF,I=1,NC
230 F(I)=F(I)/SUMF
240 GR(I)=1.
250 XF: CØNTINUE
260 PHI = 0.5
270 CALL GAMMA (A,C,E,GE,NC)
280 CALK: DØ KVALUE,I=1,NC
290 KVALUE: EK(I)=GR(I)/GE(I)
300 IF(PHI-1.)NEXT1,SKIP,NEXT1
310 NEXT1: DØ ECALC,I=1,NC
320 ECALC: E(I)=EK(I)*R(I)
330 CALL GAMMA(A,C,E,GE,NC)
340 ERRK = 0.
350 DØ KERR,I=1,NC
360 ERRK = ERRK+(EK(I)-GR(I)/GE(I))↑2
370 KERR: EK(I)=GR(I)/GE(I)
380 IF(0.00001-KERR)ECALC
390 SKIP: SUM=DRIV=ERR=0.
400 DØ RCALC,I=1,NC
410 RN=F(I)/((EK(I)-1.)*PHI+1.)
420 ERR = ERR+(RN-R(I))↑2
430 R(I)=RN
440 SUM=SUM+RN
450 RCALC: DRIV = DRIV+(EK(I)-1.)*F(I)/(((EK(I)-1.)*PHI+1.)↑2)
460 IF(ERR-0.00001)TESTSM
470 CALL GAMMA(A,C,R,GR,NC)
480 GØ TØ CALK
490 TESTSM: PSI=SUM-1.
500 IF(ABS(PSI)-0.00001)CØNV
510 IF(PHI-1.)NEXT2,NEXT3,NEXT2
520 NEXT2: DELTA = PSI/DRIV
530 TESDEL: IF(PHI+DELTA)REDUCE
540 PHI = PHI+ DELTA
550 GØ TØ CALK
560 NEXT3: IF(PSI)NEXT4
570 IF(DRIV)NEXT2,NEXT5,NEXT5
580 NEXT4: IF(DRIV)NEXT5,NEXT5,NEXT2
590 NEXT5: PRINT "NØ SØLUTIØN"
600 REWIND; GØ TØ STRT1
610 CØNV: PRINT "FRACTIØN IN EXTRACT =",PHI
620 CALLWF(F,WM,NC);CALL WF(E,WM,NC);CALL WF(R,WM,NC)
630 DØ ØUTPUT, I=1,NC
640 ØUTPUT: PRINT,F(I),E(I),R(I)
650 REWIND; GØ TØ STRT1
660 REDUCE: DELTA=0.5*DELTA
670 GØ TØ TESDEL
680 SUBRØUTINE GAMMA(A,C,F,G,NC)
690 DIMENSIØN A(10,10),C(10,10),F(10),X(10),G(10)
700 SUM = 0.
710 DØ TØTALF,I=1,NC
720 TØTALF: SUM = SUM + F(I)
730 DØ XCALC,I=1,NC
740 XCALC: X(I)=F(I)/SUM
750 DØ G2,I=1,NC
760 AG=BG=CG=0.
770 DØ G1,J=1,NC
```

Figure 6.13 Program for solution of Example 6-4.

```
780 AG = AG+A(I,J)*X(J)
790 BG=BG+A(J,I)*X(J)
800 G1: CG=CG+C(I,J)*X(J)
810 IF (X(I)-1.) NEXT,PURE,NEXT
820 NEXT: CØNTINUE
830 Z1=AG*X(I)/(AG*X(I)+BG*(1.-X(I)))
840 Z2=1.-Z1
850 AG=AG/(1.-X(I));CG=CG/(1.-X(I))
860 G(I)=EXP(AG*Z2↑2*(1.+CG*(Z1↑2-2.*Z1/3.)))
950 GØ TØ G2
960 PURE: G(I)=1.
970 G2: CØNTINUE
980 RETURN
990 SUBRØUTINE WF(X,WM,NC)
1000 DIMENSIØN X(10),WM(10)     wt fraction
1010 SUM=0.
1020 DØ ADD,I=1,NC
1030 ADD: SUM=SUM+X(I)*WM(I)
1040 DØ WTF,I=1,NC
1050 WTF: X(I)=WM(I)*X(I)/SUM
1060 RETURN
1070 $DATA
```

Figure 6.13 (*Continued*).

isobutyl ketone may be assumed to form an ideal binary solution. Three-suffix Van Laar equations may be used to represent the other two binaries. For acetone–water the Van Laar constants obtained by extrapolation of vapor–liquid equilibrium data are 2.07 and 1.939, respectively. The Van Laar constants calculated from mutual solubility are 5.61 and 2.55 for the methyl isobutyl ketone–water system.

Solution. A computer program for the solution of this problem on a G.E. time-sharing computer is shown in Figure 6.13. The correspondence between the FORTRAN variables and symbols of Equations 6-22 through 6-30 are as follows:

$$F(I) = x_i^F,$$
$$E(I) = x_i^I,$$
$$R(I) = x_i^{II},$$
$$PSI = \psi,$$
$$PHI = \phi.$$

The file of input information for the statements of lines 140 and 160 is as follows:

```
1080 $DATA
1090 1 1 3 0
1100 6    1 2 .000001 0.      2 1 .000001 0.
1110         1 3 2.07 0.      3 1 1.939 0.
1120         2 3 5.61 0.      3 2 2.55 0.
1140    58 100 18
```

6.5 Process Calculations

The program was run with a variety of values of x_1^F (acetone) to obtain the tie-lines and two-phase envelope shown in Figure 6.14.

These results are fairly typical of the results normally obtained by this method of estimating activity coefficients; the tie-lines are fairly reliable, but the predicted two-phase region is too large compared with experimental data.

6-46

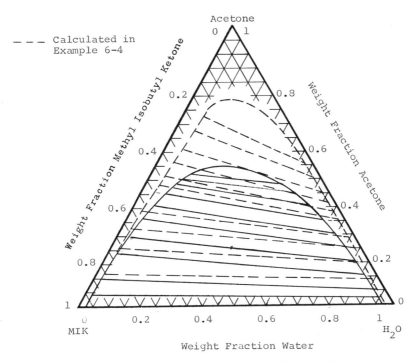

Figure 6.14 Comparison of calculated and experimental ternary diagram for acetone–methyl isobutyl ketone–water at 25°C.

A similar comparison of calculated and experimental data is shown in Figure 6.15 for aniline–n-heptane–methylcyclohexane. In this case n-heptane–methylcyclohexane was assumed to be ideal and Van Laar constants for the other binaries were calculated by using mutual solubilities. For this type of diagram the two-phase envelope as well as the tie-lines are predicted satisfactorily.

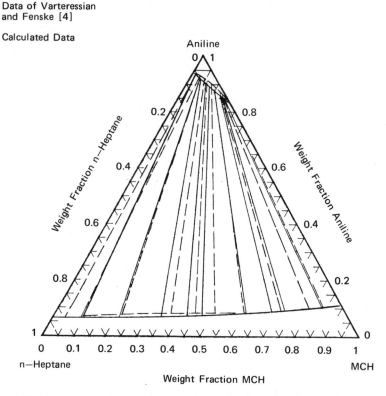

Figure 6.15 Comparison of experimental and calculated ternary diagram for aniline–*n*-heptane–methylcyclohexane at 25°C.

Three Liquid Phases. The existence of three liquid phases at equilibrium is characterized by the existence of three distinct compositions satisfying Equations 6-23. Numerical iterative procedures are required, since no analytical solution can be found. It is customary to assume that the feed of composition x_i^F splits into three phases having compositions x_i^I, x_i^{II}, and x_i^{III}, with ϕ_1 and ϕ_2 as the fraction of the feed appearing in phases I and II, respectively. A material balance gives

$$x_i^F = \phi_1 x_i^I + \phi_2 x_i^{II} + (1 - \phi_1 - \phi_2)x_i^{III}. \tag{6-31}$$

From Equations 6-23

$$x_i^I = \frac{\gamma_i^{III}}{\gamma_i^I} x_i^{III}$$

$$x_i^{II} = \frac{\gamma_i^{III}}{\gamma_i^{II}} x_i^{III}. \tag{6-32}$$

6.5 Process Calculations

Equations 6-32, substituted into Equation 6-31, give, when solved for x_i^{III}:

$$x_i^{III} = \frac{x_i^F}{[(\gamma_i^{III}/\gamma_i^{I}) - 1]\phi_1 + [(\gamma_i^{III}/\gamma_i^{II}) - 1]\phi_2 + 1}. \tag{6-33}$$

The requirement that the mole fractions in each phase must sum to unity is met if we use the alternative criteria:

$$\begin{aligned} \psi_1 &= \sum (x_i^{III} - x_i^{I}) = 0, \\ \psi_2 &= \sum (x_i^{III} - x_i^{II}) = 0. \end{aligned} \tag{6-34}$$

The simultaneous solutions of Equations 6-32, 6-33, and 6-34 are found as follows:

1. Assume initial compositions and values of ϕ_1 and ϕ_2.
2. Calculate the γ terms for each phase by using an appropriate activity-coefficient equation.
3. Calculate x_i^{III} from Equation 6-33 and x_i^{I} and x_i^{II} from Equation 6-32.
4. Calculate ψ_1 and ψ_2.
5. If ψ_1 and ψ_2 are not sufficiently close to zero, adjust ϕ_1 and ϕ_2, and, using the new calculated compositions, repeat from step 2.

The adjustment of the values of ϕ_1 and ϕ_2 can often be made successfully by the Newton–Raphson procedure. A differential change in ϕ_1 and ϕ_2 produces the following changes in ψ_1 and ψ_2:

$$\begin{aligned} \Delta\psi_1 &\simeq \frac{\partial\psi_1}{\partial\phi_1}\Delta\phi_1 + \frac{\partial\psi_1}{\partial\phi_2}\Delta\phi_2, \\ \Delta\psi_2 &\simeq \frac{\partial\psi_2}{\partial\phi_1}\Delta\phi_1 + \frac{\partial\psi_2}{\partial\phi_2}\Delta\phi_2. \end{aligned} \tag{6-35}$$

Since we wish to make ψ_1 and ψ_2 both zero, we set $\Delta\psi_1 = -\psi_1$ and $\Delta\psi_2 = -\psi_2$ and solve Equations 6-35 simultaneously for $\Delta\phi_1$ and $\Delta\phi_2$. Since Equations 6-35 are only an approximation, this will not give us the correct values of ϕ_1 and ϕ_2, but should produce a better set to use in the next iteration. The values of the partial derivatives, if we ignore the dependence of the γ terms on the ϕ terms, are evaluated as follows:

$$\frac{\partial\psi_1}{\partial\phi_1} = \sum_i \frac{(x_i^{III} - x_i^{I})[1 - (\gamma_i^{III}/\gamma_i^{I})]}{B_i},$$

$$\frac{\partial\psi_1}{\partial\phi_2} = \sum_i \frac{(x_i^{III} - x_i^{I})[1 - (\gamma_i^{III}/\gamma_i^{II})]}{B_i},$$

$$\frac{\partial\psi_2}{\partial\phi_1} = \sum_i \frac{(x_i^{III} - x_i^{II})[1 - (\gamma_i^{III}/\gamma_i^{I})]}{B_i},$$

$$\frac{\partial\psi_2}{\partial\phi_2} = \sum_i \frac{(x_i^{III} - x_i^{II})[1 - (\gamma_i^{III}/\gamma_i^{II})]}{B_i},$$

```
100 DIMENSION Z(10),X1(10),X2(10),X3(10),G1(10),G2(10),G3(10)
110COMMENT DIMENSION WM(10)
120 DIMENSION A(10,10),C(10,10)
130 READ,NC,(Z(I),I=1,NC)
140 READ,PHI1,PHI2
150 READ,((A(I,J),A(J,I),J=I,NC),I=1,NC)
160 READ,(X1(I),X2(I),X3(I),I=1,NC)
170      READ(WM(I),I=1,NC)
180 SZ=0.;DO NORM,I=1,NC
190 NORM: SZ=SZ+Z(I)
200 DO NORMZ,I=1,NC
210 NORMZ: Z(I)=Z(I)/SZ
220 ITER=0
230 CALC: ITER=ITER+1
240 IF(50-ITER)NOSOL
250 CALL GAMMA(A,C,X1,G1,NC);CALLGAMMA(A,C,X2,G2,NC);CALLGAMMA(A,C,
260 + X3,G3,NC)
270 PSI1=PSI2=PSI12=PSI11=PSI21=PSI22=0.
280 DO SUMS,I=1,NC
290 DEN = PHI1*(G3(I)/G1(I)-1.)+PHI2*(G3(I)/G2(I)-1.)+1.
300 X3(I)=Z(I)/DEN;X1(I)=G3(I)*X3(I)/G1(I);X2(I)=X3(I)*G3(I)/G2(I)
310 PSI1=PSI1+X3(I)-X2(I);PSI2=PSI2+X3(I)-X1(I)
320 PSI11=PSI11+(X3(I)-X2(I))*(1.-G3(I)/G1(I))/DEN
330 PSI12=PSI12+(X3(I)-X2(I))*(1.-G3(I)/G2(I))/DEN
340 PSI21=PSI21+(X3(I)-X1(I))*(1.-G3(I)/G1(I))/DEN
350 PSI22=PSI22+(X3(I)-X1(I))*(1.-G3(I)/G2(I))/DEN
360 SUMS: CONTINUE
370 IF(ABS(PSI1)+ABS(PSI2)-0.00001)FINISH
380 DEN=PSI11*PSI22-PSI12*PSI21
390 DEL1=(PSI2*PSI12-PSI1*PSI22)/DEN
400 DEL2=(PSI1*PSI21-PSI2*PSI11)/DEN
410 CK1: IF(-PHI1-DEL1)CK2
420 DEL1=0.5*DEL1;GO TO CK1
430 CK2: IF(-PHI2-DEL2)CK3
440 DEL2=0.5*DEL2;GO TO CK2
450 CK3: IF(PHI1+DEL1-1.)CK4
460 DEL1=0.5*DEL1; GO TO CK3
470 CK4: IF(PHI2+DEL2-1.)CK5
480 DEL2=0.5*DEL2;GO TO CK4
490 CK5: IF(PHI1+PHI2+DEL1+DEL2-1.)PASSED
500 DEL1=0.5*DEL1;DEL2=0.5*DEL2;GO TO CK5
510 PASSED: PHI1=PHI1+DEL1;PHI2=PHI2+DEL2;GO TO CALC
520 NOSOL: PRINT "NO SOLUTION,EXCESSIVE ITERATIONS"
530     CALL WF(Z,WM,NC);CALLWF(X1,WM,NC);CALL WF(X2,WM,NC)
540     CALL WF(X3,WM,NC)
550 FINISH: PRINT"FRACTIONS IN PHASES 1 AND 2",PHI1,PHI2
560 PRINT"      FEED      PHASE1      PHASE2      PHASE3"
570 DO MF,I=1,NC
580 MF: PRINT,Z(I),X1(I),X2(I),X3(I)
590 STOP

TRIFAZ CONTINUED

600 SUBROUTINE GAMMA(A,C,F,G,NC)
610 DIMENSION A(10,10),C(10,10),F(10),X(10),G(10)
620 SUM = 0.
630 DO TOTALF,I=1,NC
640 TOTALF: SUM = SUM + F(I)
650 DO XCALC,I=1,NC
660 XCALC: X(I)=F(I)/SUM
670 DO G2,I=1,NC
680 AG=BG=CG=0.
690 DO G1,J=1,NC
700 AG = AG+A(I,J)*X(J)
710 BG=BG+A(J,I)*X(J)
720 G1: CG=CG+C(I,J)*X(J)
730 IF (X(I)-1.) NEXT,PURE,NEXT
740 NEXT: CONTINUE
750 Z1=AG*X(I)/(AG*X(I)+BG*(1.-X(I)))
```

Figure 6.16 Solution of equilibrium equations for three liquid phases.

6.5 Process Calculations

```
760 Z2=1.-Z1
770 AG=AG/(1.-X(I));CG=CG/(1.-X(I))
780 G(I)=EXP(AG*Z2↑2*(1.+CG*(Z1↑2-2.*Z1/3.)))
790 G0 T0 G2
800 PURE: G(I)=1.
810 G2: C0NTINUE
820 RETURN
830 SUBR0UTINE WF(X,WM,NC)
840 DIMENSI0N X(10),WM(10)
850 SUM=0.
860 D0 ADD,I=1,NC
870 ADD: SUM=SUM+X(I)*WM(I)
880 D0 WTF,I=1,NC
890 WTF: X(I)=WM(I)*X(I)/SUM
900 RETURN
910 $DATA
920 3 0.3 0.3 0.4
930 0.3 0.3

940 0. 0.     2.878 4.68    1.593 4.588
950 0. 0.     3.16 1.496    0. 0.
960 1. 0. 0.   0. 1. 0.     0. 0. 1.

970   61.04    186.33    62.07


FRACTI0NS IN PHASES 1 AND 2      .1367       .3925
       FEED    PHASE1    PHASE2    PHASE3
        .30     .981      .0957     .2726
        .30     .0086     .72       .0345
        .40     .0104     .1844     .6929
```

Figure 6.16 (*Continued*).

where

$$B_i = \left(\frac{\gamma_i^{III}}{\gamma_i^I} - 1\right)\phi_1 + \left(\frac{\gamma_i^{III}}{\gamma_i^{II}} - 1\right)\phi_2 + 1.$$

Figure 6.16 gives a FORTRAN program and solution of a three-liquid-phase problem by the methods described. The example is the calculation of the compositions of the three liquid phases in the glycol–lauryl alcohol–nitromethane system at 22°C. The constants for the binary Van Laar equations were obtained from the mutual solubilities, as indicated on the plots given by Francis [5]. Figure 6.17 shows the calculated three-phase compositions compared with the phase diagram reported by Francis. Agreement is not as good as desired, but the calculated results are roughly correct.

6.5.3 Multicomponent Systems

The extension of ternary calculations to more than three components is straightforward, although geometric representation and visualization are more difficult. The programs used in illustrating the ternary calculations are

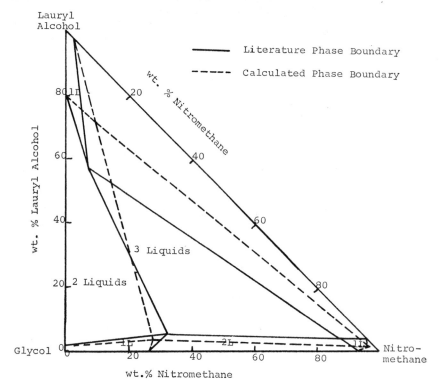

Figure 6.17 Three liquid phases in equilibrium—calculated and experimental diagram.

written for any number of components, although the examples are for three components. The number of liquid phases that can exist may be as great as the number of components, although in most practical applications only two are usually considered and the existence of a third is viewed with alarm.

6.5.4 Approximate Calculations

In many practical applications of liquid–liquid equilibria it is not necessary to use the rigorous computations outlined in the preceding sections, especially in the preliminary stages of process design. The most extensive application of liquid–liquid equilibria in the chemical industry is in liquid extraction, and the ease of separation is indicated by the quantity

$$\beta_{ij} = \frac{x_i^{\text{I}} x_j^{\text{II}}}{x_i^{\text{II}} x_j^{\text{I}}}. \tag{6-36}$$

6.5 Process Calculations

The selectivity β is entirely analogous to the relative volatility α used in vapor–liquid equilibrium calculations. A value of $\beta = 1$ indicates an impossible separation, analogous to azeotrope formation, and the ease of separation is directly proportional to $|\ln \beta|$. In comparing the separating ability of various solvents, it is sufficient to know the maximum value of β for the system studied. This normally occurs if we assume infinite dilution for the components to be separated, i and j, and complete immiscibility of the primary solvent components of phases I and II. The value of β can be expressed in terms of activity coefficients as

$$\beta_{ij} = \left(\frac{\gamma_j}{\gamma_i}\right)^{\mathrm{I}} \left(\frac{\gamma_i}{\gamma_j}\right)^{\mathrm{II}}, \tag{6-37}$$

or, at infinite dilution,

$$\beta_{ij}^{\infty} = \left(\frac{\gamma_j^{\infty}}{\gamma_i^{\infty}}\right)^{\mathrm{I}} \left(\frac{\gamma_i^{\infty}}{\gamma_j^{\infty}}\right)^{\mathrm{II}}. \tag{6-38}$$

The following example illustrates this procedure:

Example 6-5

Estimate the selectivity of diethylene glycol (phase I) for benzene (component 1) in benzene–heptane (phase II) mixtures at 75°C. Use Tables 4.1 and 4.2 to estimate the γ^{∞} value for benzene (component 1) and heptane (component 2).

Solution. The value of $(\gamma_1^{\infty})^{\mathrm{I}}$ is estimated by including the appropriate terms from Table 4.1, considering benzene as nonpolar and diethylene glycol as polar and associating.

The pertinent parameters are as follows:†

$$v_2 = \frac{106.16}{1.14318 - (0.0016 \times 75)} = 103.75 \text{ cm}^3/\text{g-mole},$$

$$v_1 = \frac{78.11}{0.89964 - (0.001074 \times 75)} = 95.36 \text{ cm}^3/\text{g-mole};$$

$$\delta_2^2 = \frac{2.303 \times 1.987 \times 1951.9}{103.75} \left(\frac{348}{118.2 + 75}\right)^2 - \frac{1.987 \times 348}{103.75}$$

$$= 272.66 \text{ cal/cm}^3,$$

$$\delta_2 = 16.512.$$

† See the following equations for the definitions of some of the parameters used here: δ^2, Equation 4-117; λ^2 Equation 4-119; K, Equation 4-121; ξ^2 Equation 4-120; τ^2, Equation 4-122.

Similarly

$$\delta_1^2 = 73.19,$$
$$\delta_1 = 8.555;$$
$$\lambda_2^2 = \frac{2.303 \times 1.987 \times 966.509}{103.75}\left(\frac{348}{75 + 95.127}\right)^2 - \frac{1.987 \times 348}{103.75}$$
$$= 171.705 \text{ cal/cm}^3,$$
$$\lambda_2 = 13.104,$$
$$\lambda_1^2 = \delta_1^2 = 73.19,$$
$$\lambda_1 = \delta_1 = 8.555;$$
$$K_2 = \exp\left(1 - \frac{3.0081}{1.987} - \frac{187.705}{1.987 \times 348}\right) = 0.78471;$$
$$\xi_2^2 = -\frac{187.705}{103.75}\left[\frac{0.78471 - \ln(1.78471)}{0.78471}\right] = -0.474;$$
$$\tau_2^2 = 272.66 - 171.705 + 0.474 = 101.429;$$
$$\Delta g_2 = \Delta h_2 - T\Delta s_2$$
$$= 187.705 - (348 \times 3.0081) = -859.114 \text{ cal/g-mole}.$$

From Table 4.1 we have the following appropriate terms for $\ln(\gamma_1^\infty)^I$:

Athermal mixing, component 1:

$$1 + \ln \frac{95.36}{103.75} = 0.91568.$$

Athermal mixing, component 2:

$$-\frac{95.36}{103.75}\left\{\frac{1}{1.78471} + \frac{\ln(0.78471)}{0.78471}\left[\ln\left(1.78471 - \frac{0.78471}{1.78471}\right)\right]\right\} = -0.43083.$$

Free energy of association bonds broken:

$$\frac{859.114}{1.987 \times 348}\left[\frac{\ln(1.78471)}{0.78471} - \frac{1}{0.78471}\right]\frac{95.36}{103.75} = 0.20313.$$

Heat of mixing:

$$\frac{95.36}{1.987 \times 348}\{(13.104 - 8.55)^2 + 101.429[0.020295$$
$$- 0.007543(13.104 - 8.555)^2]\} = 0.95429.$$

$$\ln(\gamma_1^\infty)^I = 0.91568 - 0.43083 + 0.20313 + 0.95429 = 1.6423,$$
$$(\gamma_1^\infty)^I = 5.167.$$

6.5 Process Calculations

In similar manner we calculate

$$(\gamma_3^\infty)^\mathrm{I} = 91.048,$$
$$(\gamma_1^\infty)^\mathrm{II} = 1.293,$$
$$(\gamma_3^\infty)^\mathrm{II} = 1,$$

since phase II is heptane,

$$\beta_{13} = \frac{1.293}{5.167}\frac{91.048}{1} = 22.78.$$

The experimental data of Johnson and Francis [6] indicate a value of 16.7 for β at $x_{AB} = 0.013$.

In order to compare several solvents for selectivity, one would merely repeat Example 6-5 with several other solvents. The selection is not necessarily the one with the highest β value, however, since high selectivity often means low capacity of the solvent for the extracted component. A rough estimate of maximum capacity can be made by assuming the solubility to be given by

$$(x_i)_{\max}^\mathrm{I} \simeq \frac{1}{(\gamma_i^\infty)^\mathrm{I}}.$$

Thus a solvent is chosen on the economic basis of a combination of factors, the desirable factors being as follows:

1. High β value.
2. Reasonably high $(x_i)_{\max}^\mathrm{I}$.
3. Low cost.

Another useful approximate calculation is the estimation of the distribution of components between two solvents, assumed to be completely immiscible. The immiscibility assumption effectively reduces the number of components to be considered in Equations 6-1, since each solvent is considered to be present in only one phase. To calculate a distribution coefficient merely assume a composition of one phase and calculate the mole fraction of each distributed component by

$$x_i^\mathrm{II} = \frac{\gamma_i^\mathrm{I}}{\gamma_i^\mathrm{II}} x_i^\mathrm{I}. \tag{6-39}$$

In general γ_i^I is fixed by the assumed composition of phase I and γ_i^II must be found by successive approximations of Equation 6-39 and the x_i^II values calculated. Sometimes, in practical calculations, it is feasible to assume phase II to be an ideal solution, in which case $\gamma_i^\mathrm{II} = 1$ and Equation 6-39 is solved directly. After distributed component mole fractions are calculated, the solvent mole fraction can be estimated by difference. The following example illustrates the estimation of component distributions by this method:

Example 6-6

Calculate the distribution of acetone between water and methyl isobutyl ketone at 25°C, assuming the organic phase to be an ideal solution and the aqueous phase to consist of the binary acetone–water system, with Van Laar constants of 2.07 and 1.939, respectively. Make the calculations at 0.01 mole fraction acetone increments of the aqueous phase and express the results in weight fractions.

Solution. Equation 6-39, for the conditions of this problem, can be expressed for acetone as

$$x^{II} = \gamma^I x^I.$$

For this binary system the Van Laar equation simplifies to

$$\ln \gamma = 2.07(1 - Z)^2,$$

where

$$Z = \frac{2.07x}{2.07x + 1.939(1 - x)}.$$

Thus for $x = 0.01$

$$Z = \frac{0.0207}{0.0207 + 1.920} = 0.01066,$$

$$\ln \gamma = 2.07 \times 0.98934^2 = 2.048,$$

$$\gamma = 7.58,$$

$$x^{II} = 7.58 \times 0.01 = 0.0758.$$

Converting to weight fractions, we obtain

$$w^I = \frac{58 \times 0.01}{(58 \times 0.01) + (18 \times 0.99)} = 0.0315,$$

$$w^{II} = \frac{58 \times 0.0758}{(58 \times 0.0758) + (100 \times 0.9242)} = 0.0454.$$

The results of calculations at other compositions are as follows:

Mole Fraction		Weight Fraction	
Aqueous	Organic	Aqueous	Organic
.01	.0758	.0315	.0454
.02	.1452	.0617	.0897
.03	.2087	.0906	.1327
.04	.2668	.1184	.1742
.05	.3198	.145	.2142
.06	.3682	.1706	.2526
.07	.4124	.1952	.2893
.08	.4527	.2189	.3242
.09	.4895	.2417	.3574
.10	.5229	.2636	.3887
.11	.5534	.2848	.4181
.12	.581	.3053	.4458
.13	.6062	.325	.4717
.14	.629	.3441	.4958
.15	.6496	.3625	.5181
.16	.6683	.3803	.5388
.17	.6851	.3976	.5579
.18	.7003	.4143	.5755
.19	.714	.4305	.5916

The results of Examples 6-6 and 6-4 are shown in Figure 6.18. This plot shows that the approximate calculation yields results that agree with experimental values about as well as the more rigorous calculation. Both agree with the experimental values well enough for design calculations for extraction processes until the *plait point* (point at which the two liquid phases merge into one phase) is approached. In general one avoids operation in the plait-point region because selectivity is poor in this region and there is danger of losing one of the phases. Since neither method predicts the plait point with sufficient accuracy, a practical approach is to use the approximate calculation coupled with some experimental measurement of the miscibility envelope to guide the process designer regarding the concentration limits of the method.

6.5.5 Process Applications

The applications of liquid–liquid equilibria parallel those of vapor–liquid equilibria very closely. Liquid–liquid extraction, in terms of process applications, is directly analogous to distillation. Thus the discussion of Section 5.5.5 is directly applicable to applications of liquid–liquid equilibria if one merely considers the less dense of the two liquid phases as corresponding to the vapor phase in the discussions of Chapter 5. Thus in liquid–liquid applications, as well as vapor–liquid applications, knowledge of the equilibrium relationships is required whether one is designing stagewise or continuous countercurrent equipment.

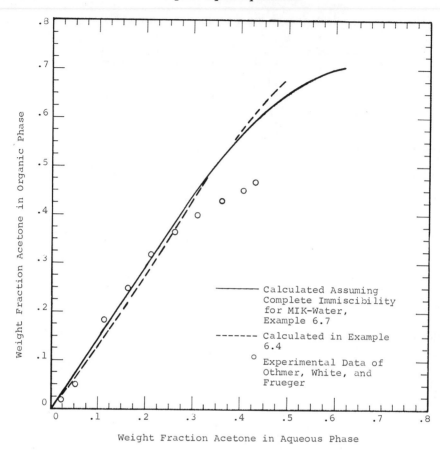

Figure 6.18 Comparison of acetone-distribution calculations between water and methyl isobutyl ketone. Solid curve: calculated assuming complete immiscibility for methyl isobutyl ketone–water, Example 6-6; dashed curve: calculated in Example 6-4; circles: experimental data of Othmer, White, and Frueger [3].

REFERENCES

[1] H. C. Carlson and A. P. Colburn, *Ind. Eng. Chem.*, **34**, 581 (1942).
[2] C. Black, *Ind. Eng. Chem.*, **50**, 403 (1958).
[3] D. F. Othmer, R. E. White, and E. Frueger, *Ind. Eng. Chem.*, **33**, 1240 (1941).
[4] K. A. Varteressian and M. R. Fenske, *Ind. Eng. Chem.*, **29**, 270 (1937).
[5] A. W. Francis, *J. Phys. Chem.*, **60**, 20 (1956).
[6] G. C. Johnson and A. W. Francis, *Ind. Eng. Chem.*, **46**, 1662 (1954).

PROBLEMS

1. Darwent and Winkler† give the following mutual-solubility data at 25°C:

	Mole Fraction Solubility
Aniline in n-heptane	0.067
n-Hexane in aniline	0.080
Aniline in methylcyclopentane	0.143
Methylcyclopentane in aniline	0.242

Assume that n-hexane and methylcyclopentane form an ideal solution. Calculate the ternary phase diagram for the aniline–n-hexane–methylcyclopentane system at 25°C based on these data. Include at least five tie-lines.

2. Use infinite-dilution activity-coefficient correlations to estimate Van Laar parameters for the system methanol–n-hexane–n-heptane at 25°C. Use these values to estimate a ternary phase diagram for the system.

3. Diethylene glycol is brought into contact with a mixture containing 0.2 mole fraction n-hexane, 0.7 mole fraction n-heptane, and 0.1 mole fraction benzene. Assume diethylene glycol to be completely insoluble in the hydrocarbon phase. If equimolar quantities of solvent (diethylene glycol) and hydrocarbon are contacted at 50°C, what will be the quantities and compositions of the phases at equilibrium?

† *J. Phys. Chem.*, **47**, 442 (1943).

Chapter 7

SOLID–LIQUID EQUILIBRIUM

7.1 INTRODUCTION

Solid–liquid equilibrium is, in many respects, much more difficult to describe quantitatively than either vapor–liquid or liquid–liquid equilibrium. The reasons for the difficulty are both mathematical and experimental. The mathematical difficulty lies in the fact that activity coefficients in both phases must be considered and there is no way to estimate either of them independently. The experimental difficulty lies in the extreme difficulty of obtaining a true equilibrium composition of the solid phase. Solid-phase material is rarely of the same composition throughout because of the very low diffusivity of molecules in the solid state, and solid samples have portions of the liquid phase clinging to the surface, held in pores, or occluded within the solid material.

The use of solid–liquid processing is widespread throughout industry, however. Almost every solid product produced by the process industries has at some time been crystallized out of the liquid phase, and the most ultrapure materials ever produced are produced by the zone-melting technique. Crystallization processes have extremely high selectivity, due to very high activity coefficients of solid solutes; but some of the advantage is lost by the amount of liquid phase that is inevitably attached to, or occluded in, the solid material. Thus in solid–liquid equilibrium we have potentially the most selective of all separations methods combined with the least developed quantitative techniques for designing the processes.

7.2 TYPES OF PHASE DIAGRAM

Binary solid–liquid phase diagrams are almost always isobaric plots of freezing and melting temperatures versus composition. They are very similar

7.2 Types of Phase Diagram

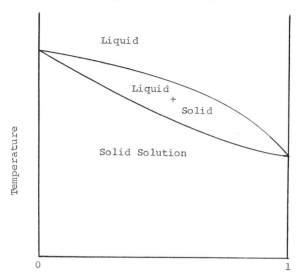

Figure 7.1 Ideal-solution phase diagram.

in appearance to the vapor–liquid diagrams, with liquid regions corresponding to the vapor regions of the vapor–liquid diagrams and solid regions corresponding to the liquid regions. For example, we have in Figure 7.1 the type of diagram exhibited when both phases form ideal solutions. This type of diagram is very rare in solid–liquid equilibrium since solid solution requires the two constituents to have molecules of almost identical size and shape.

The type of diagram shown in Figure 7.2 is exhibited by systems in which both phases are completely soluble but not ideal. The "pinching" tendency may occur at either end of the composition range.

Figure 7.3 is a diagram of a eutectic (minimum freezing point) forming system in which the solid phase is a solid solution throughout the entire composition range.

Figure 7.4 is the eutectic system in which the solid components are only partially miscible. In the extreme case the horizontal line at the eutectic temperature extends virtually across the entire composition range and the single solid phase regions disappear.

Diagrams similar to Figure 7.5 are sometimes encountered. The maximum temperature is usually considered to represent the formation of a solid-state compound between the two components. Thus the diagram really represents two binary systems—component 2 with the compound to the left of the maximum point and component 1 with the compound to the right. Each of

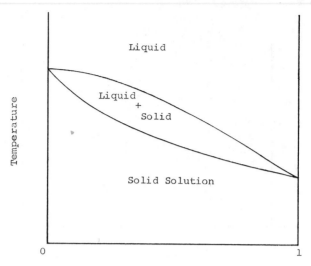

Figure 7.2 Nonideal system with complete solid solubility.

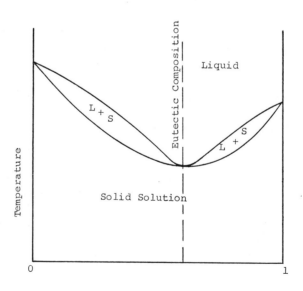

Figure 7.3 Eutectic system with complete solid solubility.

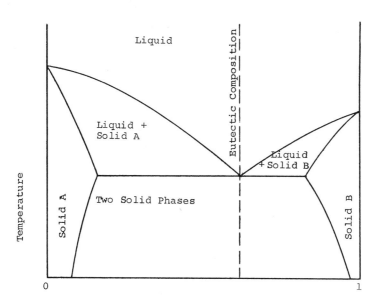

Figure 7.4 Eutectic system with limited solid solubility.

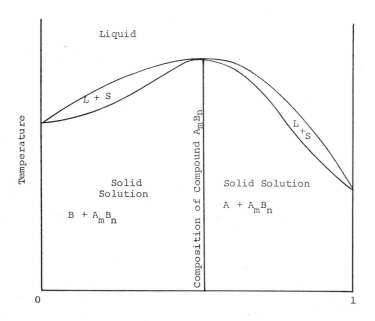

Figure 7.5 Phase diagram exhibiting complete solid solubility.

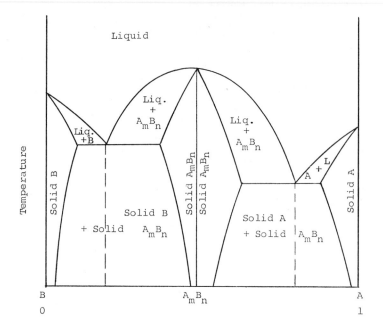

Figure 7.6 Compound and eutectic formation with limited solid solubility.

those two systems can exhibit either total or partial solubility in the solid phase; thus diagrams similar to Figure 7.6 can exist. Very complicated diagrams can exist if the compound is stable over only a portion of the composition range or if the equilibrium constant for its formation is such that appreciable quantities of both components can exist simultaneously at equilibrium with the compound.

Almost all phase diagrams presented in the literature are based on binary systems or presented as pseudobinary systems. No attempt is made to exhibit multicomponent phase diagrams here, although the same techniques that are used for ternary vapor–liquid equilibrium diagrams could be employed. The thermodynamic interpretation of these solid-liquid diagrams is quite similar to that of the vapor–liquid diagrams (see Section 5.2.7) if we adopt a correspondence between γ of the vapor–liquid case and (γ_S/γ_L) of the solid–liquid case, identify y with liquid composition and x with solid composition, and let p^* of the vapor-liquid discussion correspond to

$$\left(\frac{T^*}{T}\right)^{(C_S - C_L)/R} \exp\left(\frac{T^* - T}{RT}\right)\left(\frac{H_F}{T^*} + C_S - C_L\right).$$

7.3 Experimental Solid–Liquid Equilibrium Data

The major point of difference is that negative deviations from ideal-solution behavior are extremely rare in solid–liquid equilibrium. The eutectic corresponds in concept to the minimum boiling azeotrope. Maximum melting mixtures, however, almost always occur at a definite composition corresponding to the molecular composition of a specific compound. Thus the maximum melting composition is not exactly analogous to the high-boiling azeotrope. On the other hand, limited solid-phase solubility is analogous to limited liquid solubility; limited solubility occurs much more frequently in solids than in liquids.

7.3 EXPERIMENTAL SOLID–LIQUID EQUILIBRIUM DATA

Valid equilibrium data between a solid phase and a liquid phase are substantially more difficult to obtain than either vapor–liquid or liquid–liquid data. The two most frequently used methods consist of the following:

1. The indirect determination of the phase diagram by means of cooling and/or heating curves.
2. The direct measurement of activity coefficients in concentration cells.

7.3.1 Use of Heating and Cooling Curves

If a completely liquid solution is cooled at a uniform rate of heat removal, there will be a characteristic, relatively constant, rate of temperature drop with time. When the freezing point of the mixture is reached, however, latent heat of freezing as well as sensible heat due to temperature change must be removed; consequently, as long as the phase change is occurring, the temperature drops at a much slower rate. The resulting time-versus-temperature curve looks somewhat like that in Figure 7.7. If this type of experiment is done

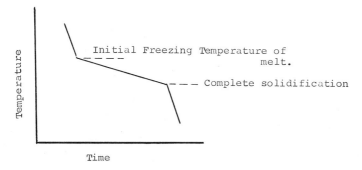

Figure 7.7 Cooling curve for obtaining freezing points.

at a number of different liquid concentrations, the freezing-point curve of the phase diagram can be determined. The complete solidification temperature is of no significance since the composition is unknown at that time.

Similarly, if a solid solution is heated at a uniform rate of heating, a fairly rapid temperature rise will be observed until the first liquid is formed, after which a lower rate of heating will occur, resulting in a temperature-versus-time curve similar to that in Figure 7.8. From a series of determinations of this type the melting point curve, or solidus, of the phase diagram can be

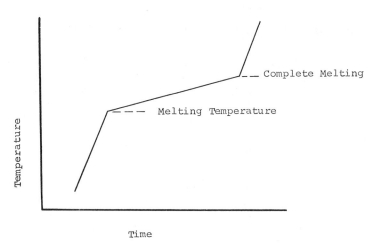

Figure 7.8 Heating curve for melting-point determination.

obtained. The corresponding equilibrium compositions at any given temperature must be obtained by the intersection of the isothermal line with the solidus and the liquidus curves of the complete phase diagram. This type of experimental data will not give individual activity coefficients, but rather the activity-coefficient ratio between the two phases, according to the following equation:

$$\frac{(\gamma_i)_S}{(\gamma_i)_L} = \frac{(x_i)_L}{(x_i)_S} \exp\left(\frac{-\Delta\mu_i^\circ}{RT}\right). \tag{7-1}$$

Aside from the necessity of cross-plotting in order to obtain the equilibrium compositions, this method suffers from the difficulty in obtaining a solid solution that is truly uniform throughout. Such a uniform solution must be obtained, however, if the data are to represent valid equilibrium data. The diffusion rate of molecules in solids is so slow that in some cases weeks,

7.3 Experimental Solid–Liquid Equilibrium Data

months, or even years of annealing are necessary to obtain uniform solutions. Unless the temperature is fairly high, even centuries may not produce a uniform solid solution.

A number of variations of this method are possible. One can plot any property that changes significantly between the two phases as a function of temperature and obtain sharp break points, similar to those obtained in the temperature-versus-time curves. Gitlesen and Motzfeldt [1], for example, used electrical conductance in determining the phase diagram of the Na_2SO_4–Na_2CO_3 system.

7.3.2 Concentration Cells

The equivalence of electrical and chemical energy can sometimes be used to obtain a direct measurement of activity coefficients. If two bars of metal, one consisting of pure metal, component i, and the other consisting of an alloy containing component i are placed in an electrolyte solution containing ions of component i, there will generally be a tendency for component i to go into the electrolyte solution at the pure bar and to be plated out on the alloy bar. A change in free energy of the total system which occurs by such a process in which ΔN moles are transferred is given by

$$\Delta F = \Delta N(\mu_i - \mu_i^\circ).$$

If an electrical potential is applied across the two bars, as shown in Figure 7.9, and just enough potential difference is applied to prevent any transfer of material from one electrode to the other, the electrical potential energy required to prevent the transfer of a quantity ΔN is given by

$$\text{electrical potential energy} = \Delta E n J \, \Delta N,$$

where n is the valence of i in the electrolyte and J is the electrical equivalent of heat.

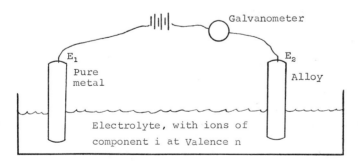

Figure 7.9 Concentration cell.

At equilibrium, or zero current flow, the free-energy change and the electrical potential energy indicated by the two equations above will be equal:

$$\mu_i - \mu_i^\circ = RT \ln \gamma_i x_i = nJ \, \Delta E,$$

or

$$\gamma_i = \frac{1}{x_i} \exp\left(\frac{nJ \, \Delta E}{RT}\right). \tag{7-2}$$

Such measurements undoubtedly represent the most accurate, and certainly the most direct, measurement of activity coefficients. The unfortunate circumstance is that it is necessary to find a suitable electrolyte that will contain the ion under consideration and be a liquid at the desired temperature.

7.3.3 Direct Measurement of Equilibrium Compositions

A method recently described by Liscom, Weinberger, and Powers [2] shows promise of more widespread use. A liquid of known composition is charged to a tube. Heat is slowly removed from the lower portion of the tube, while the liquid is stirred to maintain a uniform composition throughout, with temperature of liquid and solid cylinder height measured continually. When solidification is complete, the solid cylinder is cut into thin disks, each of which is weighed and analyzed. The initial liquid charge composition and the weights and composition of all disks formed prior to a specific disk give, by material balance, the liquid composition at the time the specific disk was formed. Thus corresponding liquid and solid equilibrium compositions are obtained. The accuracy of the method is quite sensitive to the rate of cooling. The growth conditions of the solid must be such that a uniform, reasonably flat crystallization front is maintained and that negligibly little liquid phase is occluded in the solid.

7.4 PROCESSING EXPERIMENTAL DATA

Solid–liquid phase diagrams are not expressed in the form of mathematical equations nearly as frequently as either vapor–liquid or liquid–liquid diagrams. Thus in most cases the data processing consists merely of plotting the experimental data as a phase diagram.

There are a number of problems leading to the lack of formal data-processing methods for solid–liquid equilibrium data. There is no general thermodynamic consistency test, although in some special cases conditions for thermodynamic consistency can be combined with "engineering judgment" to assess the validity of experimental data [3]. It is particularly unfortunate that no such consistency test exists since the composition uniformity of the solid phase in a heating-curve experiment is always in question.

Since the types of data that one can obtain rarely allow for independent determination of liquid-phase and solid-phase activity coefficients, but rather their ratio, there have been few attempts to determine the types of equations necessary to describe the composition dependency of the activity coefficients in the solid phase. There are individual accounts in the literature of successful use of the symmetrical and asymmetrical Van Laar equations [1], four-suffix Margules equations and Wilson equations [4], but no general conclusions.

Since the solidus curve is almost always in question, it would seem reasonable to base mathematical data processing entirely on the more reliable liquidus curve. Such an analysis has been presented [4]; separate Wilson equations were used to express liquid-phase and solid-phase activity coefficients. The constants of the Wilson equations were used as regression parameters, with $\sum [T_F(\text{calc}) - T_F(\text{exp})]^2$ as the objective function to be minimized (T_F = freezing point temperature). The procedure resulted in satisfactory calculated solidus curves; thus the method has been proven feasible, but it is not fully developed.

Undoubtedly there will be much progress in the future toward mathematical representation of solid–liquid equilibrium data. At the present, however, the use of graphical representation is the prevailing mode.

7.5 PROCESS CALCULATIONS

The process calculations desired from solid–liquid equilibrium relationships are quite analogous to those desired from vapor–liquid calculations. Given an overall composition and temperature, it is desirable to know whether the resultant material will be liquid, solid, or a mixture of both phases. If it is a mixture, it is necessary to know the quantity and composition of each phase. Since the phase-data representation is usually graphical rather than analytical, one needs only to locate the point on the phase diagram corresponding to the overall composition and temperature. If this point lies in a two-phase region, the compositions of the two phases are read at the intersections of the isotherm with the liquidus and solidus curves. A simple material balance then gives the fraction of each of the phases.

Although the method described above is the most generally used, it is not readily adaptable to multicomponent systems or to computer computations. The latter is a disadvantage only if repetitive computations are to be made on the same system.

If analytical expressions are available for the phase-equilibrium data, a set of computational procedures can be devised entirely analogous to those developed for vapor–liquid and liquid–liquid equilibrium. In many instances the mutual solubility in the solid phase is negligibly small and simplified calculation procedures can be developed by assuming the solid phase to be a pure component. Both the rigorous and simplified calculations are described.

7.5.1 Rigorous Calculations

If the activity coefficients of each phase are expressed as mathematical functions of composition, it is possible to make rigorous calculations of the usual types desired in process calculations. Given the composition of the liquid melt, the temperature at which the first solid will form and the composition of that solid can be calculated. This is the freezing-point calculation illustrated in the following example:

Example 7-1

The Wilson equations have been found to fit the experimental data of Gitlesen and Motzfeldt [1] for the Na_2CO_3–Na_2SO_4 system quite well when an appropriate temperature variation is included [4]. Calculate the freezing points of mixtures containing 0.2, 0.4, 0.6, and 0.8 mole fraction Na_2CO_3, using the Van Laar equations without temperature variations for the parameters. The infinite-dilution activity coefficients of the reference paper are to be used, as follows:

Liquid phase:
$$Na_2CO_3, \gamma_L^\infty = 3.8;$$
$$Na_2SO_4, \gamma_L^\infty = 2.5.$$

Solid phase:
$$Na_2CO_3, \gamma_S^\infty = 5.12;$$
$$Na_2SO_4, \gamma_S^\infty = 3.69.$$

The required thermodynamic properties are as follows:

Component	Melting Point (°K)	Heat of Fusion (cal/g-mole)	$(C_L - C_S)$ [cal/(g-mole)(°K)]
Na_2CO_3	1133	7300	−2
Na_2SO_4	1157	5670	4.5

Solution. The values of γ^∞ are converted to Van Laar constants:

Solid phase:
$$A_S = \ln(\gamma_{1S}^\infty) = \ln(5.12) = 1.633,$$
$$B_S = 1.305.$$

7.5 Process Calculations

Liquid phase:
$$A_L = 1.335,$$
$$B_L = 0.916.$$

With these values, the general procedure is as follows:

1. For each liquid-phase composition assume a temperature.
2. Calculate the ideal equilibrium distribution θ by Equation 3-30:

$$\theta = \frac{\gamma_L x_L}{\gamma_S x_S} = \left(\frac{T}{T^*}\right)^{(C_L - C_S)/R} \exp\left[\left(\frac{T - T^*}{RT}\right)\left(\frac{H_F}{T^*} - C_L + C_S\right)\right]$$

for each component.

3. Assume a solid composition.
4. Calculate activity coefficients for each component in each phase.
5. Calculate x_S for each component by

$$x_S = \frac{\gamma_L x_L}{\gamma_S \theta}.$$

6. Repeat steps 4 and 5 until successive x_S values agree.
7. Calculate $\sum x_{Si}$.
8. Repeat steps 3 through 7 at several temperature levels.
9. Plot $\sum x_{Si}$ and x_{Si} against T.
10. Pick the corresponding values of T and x_{Si} for which $\sum x_{Si} = 1$.

The entire procedure can be done by hand calculation, or it could all be programmed for machine computation. A combination is presented here. A program for machine computation of steps 1 through 8 is shown in Figure 7.10. The pertinent FORTRAN symbols and their meaning are as follows:

$$\text{TH1, TH2} = \theta_1, \theta_2,$$
$$\text{HF1, HF2} = H_{F1}, H_{F2},$$
$$\text{TM1, TM2} = T_1^*, T_2^*,$$
$$\text{DC1, DC2} = (C_{L1} - C_{S1}), (C_{L2} - C_{S2}),$$
$$Y = x_{L1},$$
$$X = x_{S1},$$
$$\text{G1S, G2S, G1L, G2L} = \gamma_{1S}, \gamma_{2S}, \gamma_{1L}, \gamma_{2L},$$

AS, BS, AL, BL = Van Laar constants A_{12}, A_{21} for solid and liquid phases, respectively.

Component 1 is Na_2CO_3.

```
      DIMENSION TITLE (18)
    1 READ    (5,1001) TITLE
      READ    (5,1000) A1S,B1S,A2S,B2S,A1L,B1L,A2L,B2L
      READ    (5,1000) HF1,HF2,TM1,TM2,DC1,DC2
      READ    (5,1000) NY
      WRITE   (6,1002) TITLE
      LY      = 0
    2 READ    (5,1000) Y,NT
      R       = 1.987
      LY      = LY + 1
      LT      = 0
    3 READ    (5,1000) T
      LT      = LT + 1
      X       = Y
      TH1     = (T/TM1)**(DC1/R)*EXP((T-TM1)/(R*T)*(HF1/TM1-DC1))
      TH2     = (T/TM2)**(DC2/R)*EXP((T-TM2)/(R*T)*(HF2/TM2-DC2))
      AS      = A1S + B1S/T
      BS      = A2S + B2S/T
      AL      = A1L +' B1L/T
      BL      = A2L + B2L/T
   10 ZS      = AS*X/(AS*X+BS*(1.-X))
      ZL      = AL*Y/(AL*Y+BL*(1.-Y))
      G1S     = EXP(AS*(1.-ZS)**2)
      G2S     = EXP(BS*ZS**2)
      G1L     = EXP(AL*(1.-ZL)**2)
      G2L     = EXP(BL*ZL**2)
      X1      = G1L*Y/(G1S*TH1)
      X2      = G2L*(1.-Y)/(G2S*TH2)
      IF      ( ABS(X1-X) .LT. 0.00001) GO TO 100
      X       = X1
      GO TO 10
  100 SUMX    = X1 + X2
      WRITE (6,1003)    X1,Y,T,SUMX
      IF      (LT .LT. NT) GO TO 3
      IF      (LY .LT. NY) GO TO 2
      GO TO 1
 1000 FORMAT (G1.0)
 1001 FORMAT (18A4)
 1002 FORMAT (1H1,18A4 / 8X,'X', 14X,'Y',14X,'T',11X,'SUMX')
 1003 FORMAT (2F15.5,F15.2, F15.6)
      END
```

Figure 7.10 Program for calculating solid composition from liquidus point.

The calculated results, output from the program, are shown in Figure 7.11. The plots of these results, step 9, are shown in Figure 7.12, with the final solution, step 10, shown in Table 7.1.

Table 7.1 Solution for Example 7-1

Mole Fraction Na_2CO_3 in Liquid	Mole Fraction Na_2CO_3 in Solid	Freezing Point (°K)
0.2	0.128	1132.8
0.4	0.291	1114.8
0.6	0.619	1106.8
0.8	0.840	1116.1

7.5 Process Calculations

NA2CO3 - NA2SO4 SYSTEM

X	Y	T	SUMX
0.15197	0.20000	1100.00	1.078096
0.14365	0.20000	1110.00	1.053348
0.13619	0.20000	1120.00	1.029353
0.12945	0.20000	1130.00	1.006104
0.12334	0.20000	1140.00	0.983588
0.11776	0.20000	1150.00	0.961783
0.11266	0.20000	1160.00	0.940669
0.33994	0.40000	1100.00	1.034868
0.30655	0.40000	1110.00	1.010209
0.27910	0.40000	1120.00	0.985831
0.25630	0.40000	1130.00	0.962002
0.23711	0.40000	1140.00	0.938862
0.22073	0.40000	1150.00	0.916473
0.20658	0.40000	1160.00	0.894855
0.65218	0.60000	1100.00	1.019416
0.60271	0.60000	1110.00	0.992396
0.55175	0.60000	1120.00	0.964637
0.50047	0.60000	1130.00	0.936259
0.45105	0.60000	1140.00	0.907666
0.40578	0.60000	1150.00	0.879493
0.36622	0.60000	1160.00	0.852365
0.89741	0.80000	1100.00	1.045070
0.86243	0.80000	1110.00	1.017149
0.82654	0.80000	1120.00	0.988757
0.78959	0.80000	1130.00	0.959794
0.75140	0.80000	1140.00	0.930143
0.71183	0.80000	1150.00	0.899691
0.67077	0.80000	1160.00	0.868331

Figure 7.11 Output from program of Figure 7.10.

If the composition of the solid phase is given, the temperature at which the first liquid will form and the liquid composition can be calculated. This is the melting-point calculation illustrated in Example 7-2.

Example 7-2

Calculate the melting points for each of the compositions of Example 7-1, using the same data and assumptions.

Solution. The method of solution is virtually identical with that of Example 7-1:

1. For each solid-phase composition assume a temperature.
2. Calculate θ for each component.
3. Assume a liquid composition.
4. Calculate activity coefficients for each component in each phase.
5. Calculate x_L for each component by

$$x_L = \frac{\gamma_S \theta x_S}{\gamma_L}.$$

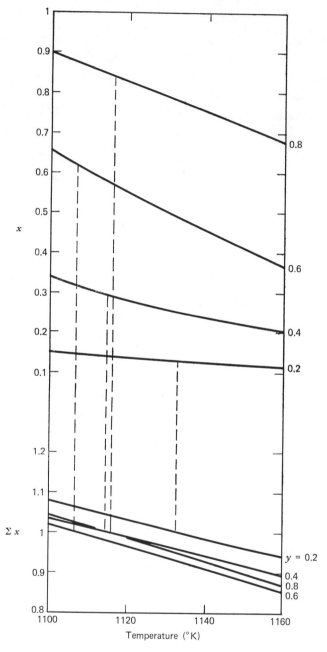

Figure 7.12 Graphical solution of Example 7-1.

7.5 Process Calculations

6. Repeat steps 4 and 5 until successive x_L values agree.
7. Calculate $\sum x_{Li}$.
8. Repeat steps 3 through 7 at several temperature levels.
9. Plot $\sum x_{Li}$ and x_{Li} against T.
10. Pick the corresponding values of T and x_{Li} for which $\sum x_{Li} = 1$.

The computer program for steps 1 through 8 is in Figure 7.13, with the output in Figure 7.14 and plotted results in Figure 7.15. The solution is shown in Table 7.2.

The results of Examples 7-1 and 7-2 are combined in Figure 7.16; the phase diagram calculated is shown compared with the experimental data. This figure is not a test of the validity of the Van Laar equations since the constants

```
         DIMENSION TITLE (18)
    1 READ    (5,1001) TITLE
      READ    (5,1000) A1S,B1S,A2S,B2S,A1L,B1L,A2L,B2L
      READ    (5,1000) HF1,HF2,TM1,TM2,DC1,DC2
      READ    (5,1000) NY
      WRITE   (6,1002) TITLE
         LY     = 0
    2 READ    (5,1000) X,NT
         R      = 1.987
         LY     = LY + 1
         LT     = 0
    3 READ    (5,1000) T
         LT     = LT + 1
         Y      = X
         TH1    = (T/TM1)**(DC1/R)*EXP((T-TM1)/(R*T)*(HF1/TM1-DC1))
         TH2    = (T/TM2)**(DC2/R)*EXP((T-TM2)/(R*T)*(HF2/TM2-DC2))
         AS     = A1S + B1S/T
         BS     = A2S + B2S/T
         AL     = A1L + B1L/T
         BL     = A2L + B2L/T
   10    ZS     = AS*X/(AS*X+BS*(1.-X))
         ZL     = AL*Y/(AL*Y+BL*(1.-Y))
         G1S    = EXP(AS*(1.-ZS)**2)
         G2S    = EXP(BS*ZS**2)
         G1L    = EXP(AL*(1.-ZL)**2)
         G2L    = EXP(BL*ZL**2)
         Y1     = G1S*TH1*X/G1L
         Y2     = G2S*TH2*(1.-X)/G2L
      IF (ABS(Y1-Y) .LT. 0.00001) GO TO 100
         Y      = Y1
      GO TO 10
  100    SUMY   = Y1+Y2
      WRITE (6,1003) X,Y1,T,SUMY
      IF  (LT .LT. NT) GO TO 3
      IF  (LY .LT. NY) GO TO 2
      GO TO 1
 1000 FORMAT (G1.0)
 1001 FORMAT (18A4)
 1002 FORMAT (1H1,18A4 / 8X,'X',14X,'Y',14X,'T',11X,'SUMY')
 1003 FORMAT (2F15.5,F15.2, F15.6)
      END
```

Figure 7.13 Program for liquid-composition computation corresponding to a solidus point.

NA2CO3 - NA2SO4 SYSTEM

X	Y	T	SUMY
0.20000	0.26194	1100.00	0.945560
0.20000	0.27984	1110.00	0.969719
0.20000	0.29898	1120.00	0.993943
0.20000	0.31936	1130.00	1.018145
0.20000	0.34094	1140.00	1.042238
0.20000	0.36366	1150.00	1.066127
0.20000	0.38740	1160.00	1.089723
0.40000	0.44218	1100.00	0.975346
0.40000	0.47273	1110.00	1.001465
0.40000	0.50364	1120.00	1.027173
0.40000	0.53462	1130.00	1.052400
0.40000	0.56542	1140.00	1.077101
0.40000	0.59587	1150.00	1.101254
0.40000	0.62582	1160.00	1.124859
0.60000	0.56518	1100.00	0.980760
0.60000	0.59823	1110.00	1.007849
0.60000	0.63071	1120.00	1.034339
0.60000	0.66253	1130.00	1.060214
0.60000	0.69358	1140.00	1.085480
0.60000	0.72385	1150.00	1.110147
0.60000	0.75331	1160.00	1.134236
0.80000	0.71318	1100.00	0.962161
0.80000	0.74588	1110.00	0.990734
0.80000	0.77765	1120.00	1.018565
0.80000	0.80851	1130.00	1.045682
0.80000	0.83848	1140.00	1.072110
0.80000	0.86759	1150.00	1.097882
0.80000	0.89588	1160.00	1.123024

Figure 7.14 Output from program of Figure 7.13.

Table 7.2 Solution for Example 7-2

Mole Fraction Na_2CO_3		Melting Point ($°K$)
Solid	Liquid	
0.2	0.302	1122.8
0.4	0.472	1109.7
0.6	0.587	1107.2
0.8	0.754	1113.3

were not found by regression of the data and temperature effects were neglected. It does indicate, however, that different results are obtained with the assumptions that were made in the examples as compared with the reference when the same γ^∞ values are used.

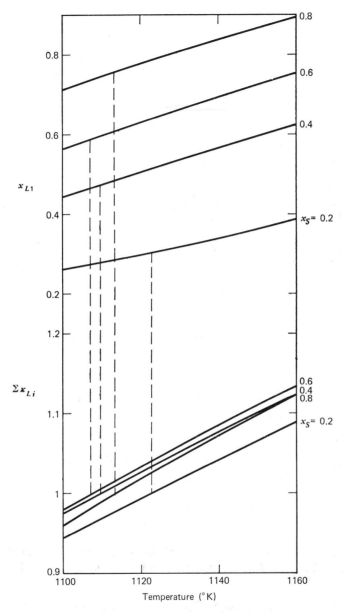

Figure 7.15 Graphical solution of Example 7-2.

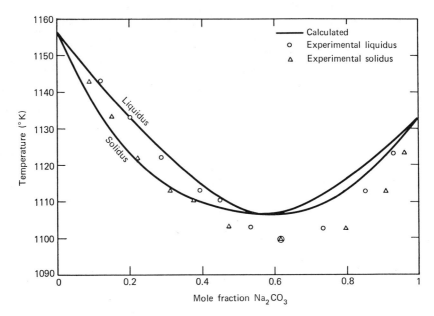

Figure 7.16 Calculated and experimental phase diagram based on Examples 7-1 and 7-2.

The following example illustrates the calculation of a phase diagram with limited solid-phase solubility:

Example 7-3

Calculate the phase diagram for a system having the same thermodynamic data as Example 7-1 except the following:

$$\ln \gamma_{1S}^\infty = 3.5,$$
$$\ln \gamma_{2S}^\infty = 2.8.$$

Solution. The same method is used as in Example 7-1. The computer output is shown in Figure 7.17; the plots, in Figure 7.18; and the phase diagram, in Figure 7.19.

The three preceding examples illustrate the calculation of the freezing and melting points, analogous to dew-point and bubble-point calculations in vapor–liquid equilibrium. A third type of calculation, analogous to the flash calculation, is the calculation of the continuous-crystallization process. In this calculation a molten feed of composition x_F is fed to a crystallization chamber

LIMITED SOLUBILITY SLE DIAGRAM

X	Y	T	SUMX
0.00974	0.10000	1100.00	1.047307
0.00942	0.10000	1110.00	1.024029
0.00912	0.10000	1120.00	1.001489
0.00884	0.10000	1130.00	0.979660
0.00857	0.10000	1140.00	0.958510
0.00832	0.10000	1150.00	0.938013
0.00808	0.10000	1160.00	0.918144
0.01679	0.20000	1080.00	1.028154
0.01618	0.20000	1090.00	1.004782
0.01561	0.20000	1100.00	0.982170
0.01508	0.20000	1110.00	0.960285
0.01457	0.20000	1120.00	0.939100
0.01410	0.20000	1130.00	0.918585
0.02495	0.30000	1040.00	1.065925
0.02390	0.30000	1050.00	1.040601
0.02291	0.30000	1060.00	1.016140
0.02200	0.30000	1070.00	0.992502
0.02115	0.30000	1080.00	0.969653
0.02035	0.30000	1090.00	0.947557
0.03580	0.40000	1000.00	1.106204
0.03396	0.40000	1010.00	1.078670
0.03228	0.40000	1020.00	1.052113
0.03075	0.40000	1030.00	1.026491
0.02934	0.40000	1040.00	1.001763
0.02804	0.40000	1050.00	0.977891
0.02684	0.40000	1060.00	0.954834
1.10748	0.50000	960.00	1.148318
1.07584	0.50000	970.00	1.121912
1.04222	0.50000	980.00	1.094602
1.00582	0.50000	990.00	1.066027
0.04148	0.50000	1000.00	1.025193
0.03922	0.50000	1010.00	0.999537
1.05497	0.60000	1000.00	1.096891
1.02167	0.60000	1010.00	1.069244
0.98529	0.60000	1020.00	1.039938
0.94413	0.60000	1030.00	1.008096
0.89436	0.60000	1040.00	0.971799
0.03649	0.60000	1050.00	0.814936
0.03479	0.60000	1060.00	0.795502
1.12543	0.70000	1000.00	1.150908
1.09791	0.70000	1010.00	1.126155
1.06912	0.70000	1020.00	1.100605
1.03867	0.70000	1030.00	1.073990
1.00596	0.70000	1040.00	1.045934
0.96999	0.70000	1050.00	1.015807
0.92884	0.70000	1060.00	0.982443
0.99807	0.80000	1070.00	1.027083
0.96342	0.80000	1080.00	0.996495
0.92381	0.80000	1090.00	0.962248
0.87483	0.80000	1100.00	0.921178
0.79474	0.80000	1110.00	0.857859
0.03464	0.80000	1120.00	0.454879
0.99906	0.90000	1100.00	1.014318
0.96680	0.90000	1110.00	0.984024
0.93043	0.90000	1120.00	0.950186
0.88687	0.90000	1130.00	0.910180
0.82547	0.90000	1140.00	0.854958

Figure 7.17 Computer output for Example 7-3.

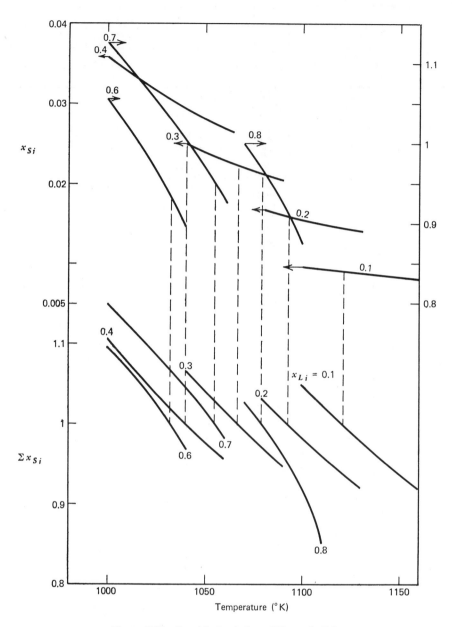

Figure 7.18 Graphical solution of Example 7-3.

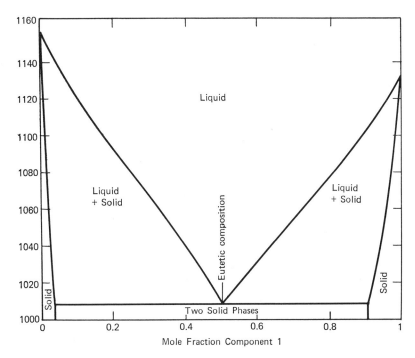

Figure 7.19 Calculated phase diagram for Example 7-3.

where it is held at a temperature T between the freezing and melting temperatures of the feed composition. If phase equilibrium is attained, there will be a solid phase and a liquid phase withdrawn at temperature T. The process calculation to be performed is to determine the fraction ϕ_S of the feed that will be withdrawn as a solid and the compositions of each phase withdrawn. The compositions of the two phases are related by material balance,

$$x_{Fi} = \phi_S x_{Si} + (1 - \phi_S) x_{Li}, \tag{7-3}$$

and by equilibrium,

$$x_{Li} = \frac{\gamma_{Si} \theta_i}{\gamma_{Li}} x_{Si}. \tag{7-4}$$

The two equation sets above and the additional restriction

$$\sum x_{Si} = \sum x_{Li} = 1 \tag{7-5}$$

are sufficient to determine ϕ_S, x_{Si}, and x_{Li}. The following example illustrates the use of a combination computer program-graphical procedure for solving the continuous-crystallization problem:

Example 7-4

Use the data of Example 7-1 to calculate the composition and quantities of the solid and liquid phases when a feed of 0.2 mole fraction Na_2CO_3 is fed into a chamber held at 1130°K.

Solution. The method of solution is as follows:

1. Assume a value for ϕ_S.
2. Initially, assume γ_{Si} and $\gamma_{Li} = 1$.
3. Solve for each x_{Si} by substituting Equation 7-4 into Equation 7-3 and rearranging to give

$$x_{Si} = \frac{x_{Fi}}{(1-\phi_S)(\gamma_{Si}/\gamma_{Fi})\theta_i + \phi_S}.$$

4. Solve Equation 7-3 for each x_{Li}:

$$x_{Li} = \frac{x_{Fi} - \phi_S x_{Si}}{1 - \phi_S}.$$

5. Calculate new values of γ_{Si} and γ_{Li} based on the calculated x_{Si} and x_{Li} values.
6. Repeat from step 3 until successive x_{Si} and x_{Li} values repeat.
7. Calculate $\sum x_{Li}$ and $\sum x_{Si}$.
8. Repeat from step 1 for a number of ϕ_S values between 0 and 1.
9. Plot $\sum x_{Si}$ and $\sum x_{Li}$ against ϕ_S.
10. Read the value of ϕ_S for which $\sum x_{Si} = 1$ and $\sum x_{Li} = 1$, between 0 and 1.

Figure 7.20 is a listing of a computer program to perform steps 1 through 7. Figure 7.21 gives the output for several values of ϕ_S, and Figure 7.22 is the plot giving the final solution:

$\phi_S = 0.327$ fraction in the solid phase,

$x_{Si} = 0.145$ mole fraction Na_2CO_3 in solid,

$x_{Li} = 0.2265$ mole fraction Na_2CO_3 in liquid.

In any continuous crystallizer a significant fraction of liquid will cling to, or will be occluded within, the solid phase. Thus the composition of the solid product will be different from that calculated in the preceding example. To correct the composition of the solid phase assume that a quantity ψ moles of liquid is carried out with each mole of solid. (The quantity ψ can only be determined by experiments on the particular crystallizer involved.) Thus each

```
      DIMENSION TITLE(18)
    1 READ  (5,1001) TITLE
      READ  (5,1000) A1S,B1S,A2S,B2S,A1L,B1L,A2L,B2L
      READ  (5,1000) HF1,HF2,TM1,TM2,DC1,DC2
      WRITE (6,1002) TITLE
    2 READ  (5,1000) XF1,T
            R     = 1.987
            XF2   = 1. - XF1
            X1A   = XF1
            X2A   = XF2
    3 READ  (5,1000) PHI
            TH1   = (T/TM1)**(DC1/R)*EXP((T-TM1)/(R*T)*(HF1/TM1-DC1))
            TH2   = (T/TM2)**(DC2/R)*EXP((T-TM2)/(R*T)*(HF2/TM2-DC2))
            AS    = A1S + B1S/T
            BS    = A2S + B2S/T
            AL    = A1L + B1L/T
            BL    = A2L + B2L/T
            G1S   = 1.
            G2S   = 1.
            G1L   = 1.
            G2L   = 1.
      GO TO 11
   10       ZS    = AS*X1/(AS*X1 + BS*X2)
            ZL    = AL*Y1/(AL*Y1 + BL*Y2)
            G1S   = EXP (AS*(1.-ZS)**2)
            G2S   = EXP (BS*ZS**2)
            G1L   = EXP (AL*(1.-ZL)**2)
            G2L   = EXP (BL*ZL**2)
   11       X1    = XF1/((1.-PHI)*G1S*TH1/G1L + PHI)
            Y1    = (XF1-PHI*X1)/(1.-PHI)
            X2    = XF2/((1.-PHI)*G2S*TH2/G2L + PHI)
            Y2    = (XF2-PHI*X2)/(1.-PHI)
      IF (ABS(X1A-X1)+ABS(X2A-X2) .LT. 0.00001) GO TO 100
            X1A   = X1
            X2A   = X2
      GO TO 10
  100       SUMX  = X1 + X2
            SUMY  = Y1 + Y2
      WRITE (6,1003) X1,Y1,SUMX,SUMY,PHI
      GO TO 3
 1000 FORMAT (G1.0)
 1001 FORMAT (18A4)
 1002 FORMAT (1H1,18A4 / 8X,'X',14X,'Y',11X,'SUMX',11X,'SUMY',12X,'PHI')
 1003 FORMAT (5F15.6)
      END
```

Figure 7.20 Program for solution of Example 7-4.

CONTINUOUS CRYSTALLIZATION PROBLEM

X	Y	SUMX	SUMY	PHI
0.129314	0.200714	1.005876	0.999941	0.010000
0.131050	0.203629	1.004977	0.999738	0.050000
0.133306	0.207410	1.003909	0.999566	0.100000
0.138131	0.215467	1.001970	0.999507	0.200000
0.143420	0.224249	1.000317	0.999864	0.300000
0.149245	0.233837	0.998978	1.000681	0.400000
0.155682	0.244318	0.997990	1.002010	0.500000
0.162815	0.255778	0.997396	1.003906	0.600000
0.170738	0.268278	0.997245	1.006428	0.700000
0.179536	0.281855	0.997593	1.009629	0.800000
0.189281	0.296469	0.998493	1.013559	0.900000
0.194521	0.304109	0.999168	1.015804	0.950000
0.198889	0.309972	0.999821	1.017744	0.990000

Figure 7.21 Output for program of Figure 7.20.

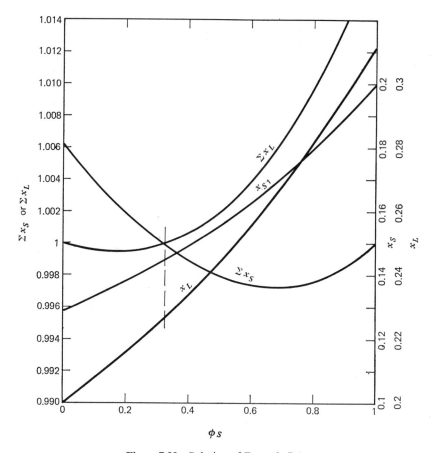

Figure 7.22 Solution of Example 7-4.

mole of solid will have associated with it $x_{Si} + \psi x_{Li}$ moles of component i and $1 + \psi$ moles total. The composition of the solid product, x_{SP}, is given by

$$x_{SPi} = \frac{x_{Si} + \psi x_{Li}}{1 + \psi}. \tag{7-6}$$

The fraction of solid product will be

$$\phi_{SP} = \phi_S(1 + \psi), \tag{7-7}$$

and the fraction of liquid product will be

$$\phi_{LP} = 1 - \phi_S(1 + \psi). \tag{7-8}$$

7.5 Process Calculations

Often, in crystallization equipment, the value of ψ is sufficiently large to affect the selectivity of the process seriously.

Example 7-5

In the continuous crystallizer of Example 7-4 each mole of solid material occluded with it 0.2 moles of liquid, which ultimately solidified in subsequent handling processes and appeared as part of the solid product. Compute the ultimate fraction of feed in the solid product and the composition of the final product.

Solution. From Equation 7-7,

$$\phi_{SP} = \phi_S(1 + \psi) = 0.327 \times 1.2 = 0.3924 \text{ moles solid product per mole feed.}$$

From Equation 7-6,

$$x_{SP1} = \frac{x_{S1} + \psi x_{L1}}{1 + \psi} = \frac{0.145 + (0.2 \times 0.2265)}{1.2} = 0.1586.$$

7.5.2 Approximate Calculations

In many processes involving crystallization of solids the mutual solubility of solid components is negligibly small. Thus each solid phase can be assumed to consist of a single pure component and the resulting calculations are greatly simplified. For the crystalline component $\gamma_S = 1$ and $x_S = 1$. Thus Equation 7-1 becomes

$$x_{Li} = \frac{\exp(\Delta \mu_i^\circ / RT)}{\gamma_{Li}}, \qquad (7\text{-}9)$$

or

$$\gamma_{Li} x_{Li} = \left(\frac{T}{T_i^*}\right)^{(C_{Li} - C_{Si})/R} \exp\left[\left(\frac{T - T_i^*}{RT}\right)\left(\frac{H_{Fi}}{T_i^*} - C_{Li} + C_{Si}\right)\right]. \quad (7\text{-}10)$$

The simplification is illustrated by the following example:

Example 7-6

The water–acetic acid binary system vapor–liquid equilibrium is described by use of the Van Laar (three-suffix) equation for liquid-phase activity coefficients. The Van Laar constants, extrapolated to the range of their freezing temperatures, are $A_{12} = 0.66952$ and $A_{21} = 0.99175$ for water as component 1. Assume complete insolubility in the solid phase and compute the following:

1. the solid–liquid phase diagram for the system and
2. the equilibrium quantities and composition of the two phases when 1 mole of 95% (mole) acetic acid is crystallized at $-20°C$.

Solid–Liquid Equilibrium

Solution for Problem 1. Equation 7-10 is used to calculate values of $\gamma_L x_L$ for each of the components at several temperatures. The required physical properties for the calculation are given by Hougen, Watson, and Ragatz [5] and the *International Critical Tables* [6] as follows:

Property	Water	Acetic Acid	Units
H_F	1436.3	2800	cal/g-mole
T^*	273.16	289.76	°K
C_L	1.008	0.483	cal/(g) (°K)
C_S	0.492	0.343	cal/(g) (°K)
ΔC	0.516	0.140	cal/(g) (°K)
	9.297	8.407	cal/(g-mole) (°K)
$\dfrac{\Delta C}{R}$	4.679	4.231	
$\dfrac{H_F}{RT^*} - \dfrac{\Delta C}{R}$	-2.033	0.632	

When the above values are substituted into Equation 7-10, the following values of $\gamma_L x_L$ are obtained:

$T(°C)$	$\gamma_{L1} x_{L1}$	$\gamma_{L2} x_{L2}$
15	1.139	0.973
0	1.0	0.75
-15	0.878	0.568
-30	0.771	0.422
-60	0.599	0.217

The values of γ_L and $\gamma_L x_L$ calculated from the Van Laar equations are listed in Table 7.3.

At each temperature desired the value of $\gamma_L x_L$ calculated by Equation 7-10 is located on the plot in Figure 7.23 of values calculated by the Van Laar equation and the corresponding liquid composition is read. The results are shown in Table 7.4. The phase diagram is shown in Figure 7.24.

Solution of Problem 2. From Figure 7.24, the composition given ($x_{L1} = 0.05$) is in the region consisting of a solid acetic acid phase in equilibrium with

7.5 Process Calculations

Table 7.3 Values of γ_L and $\gamma_L x_L$

x_{L1}	γ_{L1}	γ_{L2}	$\gamma_{L1} x_{L1}$	$\gamma_{L2} x_{L2}$
0	1.0	1.9533	0	1.0
0.05	1.0012	1.8671	0.09335	0.95114
0.1	1.0048	1.7849	0.17849	0.90432
0.2	1.0209	1.6325	0.3265	0.81672
0.3	1.0512	1.4959	0.44877	0.73584
0.5	1.1748	1.2695	0.63475	0.5874
0.7	1.4493	1.1062	0.77434	0.43479
0.8	1.6958	1.0501	0.84008	0.33916
0.9	2.0776	1.0135	0.91215	0.20770
0.95	2.3478	1.0035	0.95333	0.11739
1.0	2.6959	1.0	1.0	0

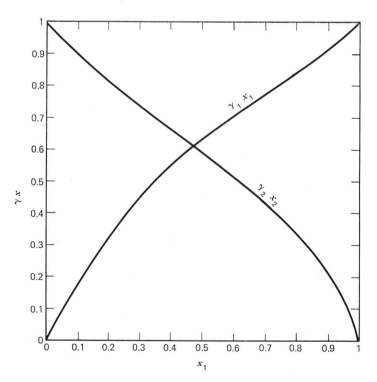

Figure 7.23 Liquid phase activities for Example 7-6.

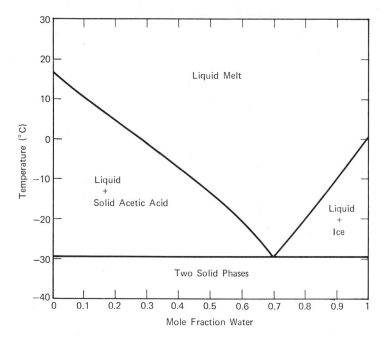

Figure 7.24 Calculated phase diagram for acetic acid–water.

Table 7.4 Calculated Equilibrium Mole Fractions

T(°C)	Liquid Mole Fraction Water in Equilibrium with Solid Acetic Acid	Liquid Mole Fraction Water in Equilibrium with Ice
15	0.027	—
0	0.28	1.0
−15	0.53	0.855
−30	0.705	0.69

liquid of composition $x_{L1} = 0.59$. The quantity in the liquid phase, ϕ_L, is given by

$$0.59\phi_L = 0.05$$

$$\phi_L = \frac{0.05}{0.59} = 0.0847 \text{ mole liquid solution.}$$

7.5 Process Calculations

The quantity of solid acetic acid is

$$1 - \phi_L = 1 - 0.0847 = 0.9153 \text{ mole.}$$

7.5.3 Process Applications

The two most frequent process applications dependent on favorable solid–liquid equilibrium are crystallization and zone refining. Crystallization is carried out by passing one or more liquid phases into a crystallizer held at a desired temperature. The solid phase separates out and a solid and liquid phase are withdrawn. Several such stages can be combined to allow solid and liquid phases to pass in cocurrent, countercurrent, or cross flow. The example problems in this chapter are crystallization applications.

Some of the purest materials ever produced have been purified by zone refining, wherein a rod of solid material is traversed a number of times by a molten zone. As the zone passes a given location in the rod, the solid melts completely; as the zone leaves the crystallized material, it is at equilibrium

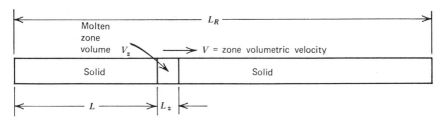

Figure 7.25 Schematic representation of zone-refining process.

with the liquid zone except for some occlusion of residual liquid. In Figure 7.25 the zone refining process is shown schematically. The rate at which a trace impurity component enters the molten zone is

$$\frac{V x_{S1} \rho_S}{M},$$

where x_{S1} is the mole fraction at point 1, ρ_S is the density of the solid, and M is the molecular weight.

The rate leaving is

$$\frac{V x_{S2} \rho_S}{M},$$

with x_S related to liquid composition by

$$x_{S2} = (1 - \phi)Kx_L + \phi x_L,$$

where ϕ is the fraction of occluded liquid and $K = \gamma_L/\gamma_S \theta$.
Material balance yields the differential equation

$$x_{S1} = x_L[(1 - \phi)K + \phi] + L_z \frac{\rho_L}{\rho_S} \frac{\partial x_L}{\partial L}.$$

With the assumption of constant K, V, ϕ, ρ_L, ρ_S, and V_z, the solution is

$$x_L = \exp\left\{-\frac{[(1 - \phi)K + \phi]\rho_S}{\rho_L \rho_S} L\right\}$$

$$\times \left\{x_{S1}^\circ + \frac{\rho_S}{\rho_L L_z} \int_0^L x_{S1} \exp\left[\frac{[(1 - \phi)K + \phi]\rho_S L}{\rho_L L_z}\right] dL\right\},$$

where x_{S1}° is the mole fraction at $L = 0$.
The refined solid product will have the composition

$$x_{S2} = [(1 - \phi)K + \phi]x_L \quad \text{when } 0 \leq L \leq L_R - L_z,$$

$$x_{S2} \simeq x_L \quad \text{when } L_R - L_z \leq L \leq L_R.$$

On the first pass of the zone, one might expect x_{S1} to be constant. On subsequent passes x_{S1} will be equal to x_{S2} of the previous pass. The assumption of constant K is usually valid in zone refining since trace impurities are being removed; thus the composition is virtually infinite dilution with respect to the component of interest.

There are numerous other applications making use of solid–liquid equilibrium properties. One familiar to all motorists is the addition of a second component (antifreeze) to lower the freezing point of a liquid, thus extending the temperature range over which it can be pumped.

REFERENCES

[1] G. Gitlesen and K. Motzfeldt, *Acta Chem. Scand.*, **18**, 488 (1964).
[2] P. W. Liscom, C. B. Weinberger, and J. E. Powers, Proc. AIChE, I. Chem. Eng. Joint Meeting, London, 1965 (1), pp. 90–104.
[3] H. R. Null, *Am. Inst. Chem. Engrs. J.*, **11**, 780 (1965).
[4] H. R. Null, *Chem. Eng. Progr. Symposium Series*, No. 81, **63**, 52 (1967).
[5] O. A. Hougen, K. M. Watson, and R. A. Ragatz, *Chemical Process Principles*, Part I, 2nd ed., John Wiley & Sons, New York, 1954.
[6] *International Critical Tables*, Vol. 3, McGraw-Hill Book Company, New York, 1927.

PROBLEMS

1. Assume water and ethylene glycol to be completely insoluble in the solid phase. Use the data of Example 7-6 for water and the following for ethylene glycol:

$H_F = 2684.6$ cal/gmole,

$T^* = -13°C$,

$\Delta C = 0$.

Calculate the liquidus curve for this binary assuming ideal liquid solution.

2. The following data for the ethylene glycol–water system are given by Ross†

Ethylene Glycol (% Wt)	Freezing Point (°C)
0	0
10	− 3.5
20	− 7.8
30	−13.7
40	−23.2
50	−36.8
60	−53.5
80	−52.0
90	−28.5

Compute liquid-phase activity coefficients from these data and plot against composition.

† *Ind. Eng. Chem.*, **46**, 601 (1954).

Chapter 8

SPECIAL TOPICS

8.1 INTRODUCTION

Proper application of the principles outlined in the preceding chapters of this book will enable the engineer to solve most of the phase-equilibrium problems he will encounter in process design. He will, however, occasionally encounter systems that will cause him to doubt all principles of any kind he has ever learned. Furthermore, the disproportionate share of time he has to spend on each such system he encounters will lead him to believe that they are far more numerous than they actually are. Actually these systems do not violate any thermodynamic principles; usually there are other forces involved besides those usually encountered in phase equilibrium alone. It would be impossible for any author to anticipate all of the special problems of this nature that will ultimately be encountered; however, it is worthwhile to discuss briefly some of the special problems that do occur. The remainder of this chapter is devoted to the discussion of a few such problems that are not often discussed in connection with phase equilibrium.

8.2 VAPOR–SOLID EQUILIBRIUM

Solids exert vapor pressures as do liquids. Consequently there are times when one is concerned with the composition of a vapor in equilibrium with a solid, or the crystallization or plating of a solid directly from the vapor. This phenomenon is exploited in the manufacture of semiconductor materials. The principles of phase equilibrium involved are identical with those discussed in Chapter 5 on vapor–liquid equilibrium, with the solid assuming the liquid role and heat of sublimation substituted for heat of vaporization. The primary difficulties are experimental. The temperature and pressure conditions of interest are usually in a realm where experimental data are difficult to obtain.

8.3 CHEMICAL REACTION

Sometimes phase-equilibrium data taken with utmost care to ensure equilibrium and eliminate sampling errors will fail to meet the requirements of the thermodynamic consistency tests. The data may be obtained by different investigators who repeatedly confirm the original data. Under such circumstances one is tempted to doubt the validity of the thermodynamic consistency tests. Usually the reason for the anomaly is an unrecognized chemical reaction. Once all the chemical species are properly accounted for, the data can be shown to be thermodynamically consistent. A book could readily be written on this subject alone; in this work we merely examine the basic principles of combined phase-chemical equilibrium and discuss some of the more frequently encountered types of chemical reaction that are most likely to go unrecognized as data are acquired.

The equations that must be satisfied between a given phase, designated phase I, and any other phase, designated M are obtained from Equations 2-21 and 2-22:

$$\frac{\gamma_i^I x_i^I}{\gamma_i^M x_i^M} = \exp\left\{\frac{[(\mu_i^\circ)^M - (\mu_i^\circ)^I]}{RT}\right\}. \tag{8-1}$$

There will be one such equation for each component in each of the phases other than phase I. Thus, if there are N_c components and N_P phases, there will be $N_c(N_P - 1)$ such equations. If, in addition, some of the components are capable of reacting, there are certain chemical reaction equilibrium equations that must be satisfied. The general equation for a reversible chemical reaction may be written

$$\sum_i a_i C_i \rightleftharpoons \sum_i b_i C_i,$$

where C_i designates the formula for component i, a_i is the coefficient of component i on the reactant side of the equation, and b_i is the product-side coefficient.

If we consider this reaction in phase M, chemical-reaction equilibrium requires that there be no change in Gibbs free energy as an increment of reaction occurs, or

$$\sum_i a_i \mu_i^M = \sum_i b_i \mu_i^M. \tag{8-2}$$

Equation 8-2 is a consequence of the principle that a system at constant pressure and temperature throughout will seek a state of minimum Gibbs free energy. By recognizing that

$$\mu_i^M = (\mu_i^\circ)^M + RT \ln(\gamma_i^M x_i^M), \tag{8-3}$$

we have

$$\sum_i a_i[(\mu_i^\circ)^M + RT \ln(\gamma_i^M x_i^M)] = \sum_i b_i[(\mu_i^\circ)^M + RT \ln(\gamma_i^M x_i^M)],$$

or

$$\sum_i RT \ln \frac{(\gamma_i^M x_i^M)^{a_i}}{(\gamma_i^M x_i^M)^{b_i}} = \sum_i (b_i - a_i)(\mu_i^\circ)^M.$$

Finally,

$$\frac{\prod_i (\gamma_i^M x_i^M)^{b_{ij}}}{\prod_i (\gamma_i^M x_i^M)^{a_{ij}}} = \exp\left[\frac{\sum_i (a_{ij} - b_{ij})(\mu_i^\circ)^M}{RT}\right], \tag{8-4}$$

where the second subscript, j, has been added to designate which of the N_R possible reactions is being described. If a given reaction can occur in any of the phases, it can occur in all the phases. Thus there are $N_R N_P$ equations of this type to be solved. Additional constraints that must be satisfied are that

$$\sum_i x_i^M = 1 \tag{8-5}$$

and that an atom balance between the resultant phases and the initial charge be met. There are N_P equations of the type of Equation 8-5. For the atom balance we may designate P_{ik} as the number of atoms of species k in the molecular structure of component i. Letting A_i represent the number of atoms (or gram-atoms or pound-atoms) of species k charged and N_M as the total molecules (or moles) present in phase M, we obtain

$$A_k = \sum_i P_{ik}\left(\sum_M N_M x_i^M\right). \tag{8-6}$$

There are N_A equations of this type, where N_A is the number of atomic species involved. For the purpose of this discussion any molecular or radical species that is not changed in any of the reactions is considered as an atom. By using this definition of atoms, we have

$$N_c = N_A + N_R. \tag{8-7}$$

The equations are summarized in Table 8.1.

Thus the total number of equations possible is $N_P(N_c + 1) + N_R(N_P - 1)$. The unknown quantities to be calculated are the N_P values of N_M and $N_P N_c$ values of x_i^M for a total of $N_P(N_c + 1)$. Thus there is a redundancy of $N_R(N_P - 1)$ equations. Actually the phase-equilibrium and chemical-equilibrium equations are not independent and $N_R(N_P - 1)$ of them can be

8.3 Chemical Reaction

Table 8.1 Equations for Systems Involving Chemical Reactions

Equation Type	Number of Equations
$\dfrac{\gamma_i^I x_i^I}{\gamma_i^M x_i^M} = \exp\left\{\dfrac{[(\mu_i^\circ)^M - (\mu_i^\circ)^I]}{RT}\right\}$	$N_c(N_P - 1)$
$\dfrac{\Pi_i(\gamma_i^M x_i^M)^{b_{ij}}}{\Pi_i(\gamma_i^M x_i^M)^{a_{ij}}} = \exp\left[\dfrac{\sum_i(a_{ij} - b_{ij})(\mu_i^\circ)^M}{RT}\right]$	$N_R N_P$
$\sum_i x_i^M = 1$	N_P
$A_k = \sum_i P_{ik}\left(\sum_M N_M x_i^M\right)$	$N_A = N_c - N_R$

eliminated as mathematical combinations of the others. This can be accomplished by specifying each chemical-reaction equation in one phase only. It is sometimes more convenient mathematically to have the liberty to choose which of the equations is to be eliminated. Equations such as Equations 8-5 and 8-6 type must never be eliminated, however.

8.3.1 Vapor-Phase Association

One of the frequent causes of apparent anomalies in phase-equilibrium data is the tendency for certain components to associate in the vapor phase. This tendency is especially strong in organic acids (especially the lower-molecular-weight species, formic, acetic, and propionic acid), in hydrogen fluoride, and to a lesser extent in some aldehydes.

The acetic acid–water system has been studied extensively, and experimenters [1–4] generally agree on the vapor–liquid equilibrium data at low pressures. Unless one is aware that the acetic acid forms dimers in the vapor phase, however, the data do not appear to be thermodynamically consistent. Pressure–volume–temperature data [5–8] on vapor-phase acetic acid have confirmed the existence of dimers and probably of associated molecules of still higher molecular weight. At low pressures, however, only the dimer is of significance, and its consideration will suffice for the present illustrative purposes. By considering the components to be (a) H_2O, (b) CH_3COOH (abbreviated AcOH), and (c) $(AcOH)_2$, we can write the equations of type of Equation 8-1 for the system:

$$\frac{y_1}{\gamma_1 x_1} = \frac{p_1^*}{\pi}; \quad \frac{y_2}{\gamma_2 x_2} = \frac{p_2^*}{\pi}; \quad \frac{y_3}{\gamma_3 x_3} = \frac{p_3^*}{\pi}, \qquad (8\text{-}8)$$

where the gas phase is considered ideal, since we are illustrating only the low-pressure case. The chemical-reaction equation involved is

$$(AcOH)_2 \rightleftharpoons 2AcOH, \tag{8-9}$$

which yields two equations of the type of Equation 8-4:

$$\frac{(\gamma_2 x_2)^2}{\gamma_3 x_3} = \exp\left\{\frac{[-2(\mu_2^\circ)^L + (\mu_3^\circ)^L]}{RT}\right\} \equiv K_L,$$
$$\frac{y_2^2}{y_3} = \exp\left\{\frac{[-2(\mu_2^\circ)^G + (\mu_3^\circ)^G]}{RT}\right\} \equiv K_G. \tag{8-10}$$

The equilibrium constant K_G varies with both temperature and pressure. To eliminate the pressure variation, we can introduce the partial-pressure equilibrium constant K_P, defined as

$$K_P = \frac{p_2^2}{p_3} = \frac{(\pi y_2)^2}{\pi y_3} = \pi K_G. \tag{8-11}$$

To illustrate the invariant quality of K_P at low pressures we differentiate Equation 8-11:

$$\left(\frac{\partial K_P}{\partial \pi}\right)_T = \pi \left(\frac{\partial K_G}{\partial \pi}\right)_T + K_G. \tag{8-12}$$

The variation of K_G can be expressed from its definition:

$$\left(\frac{\partial K_G}{\partial \pi}\right)_T = \left[-2\left(\frac{\partial \mu_2^\circ}{\partial \pi}\right)_T + \left(\frac{\partial \mu_3^\circ}{\partial \pi}\right)_T\right]\frac{K_G}{RT}$$

$$= (-2v_2 + v_3)\frac{K_G}{RT}$$

$$= \left(-2\frac{RT}{\pi} + \frac{RT}{\pi}\right)\frac{K_G}{RT} = -\frac{K_G}{\pi}. \tag{8-13}$$

Equation 8-13, substituted into Equation 8-12, yields

$$\left(\frac{\partial K_P}{\partial \pi}\right)_T = \pi\left(-\frac{K_G}{\pi}\right) + K_G = 0.$$

Thus the useful forms of Equations 8-10 become:

$$\frac{(\gamma_2 x_2)^2}{\gamma_3 x_3} = K_L; \quad \frac{\pi y_2^2}{y_3} = K_P. \tag{8-14}$$

8.3 Chemical Reaction

We now have a redundancy of one equation. For convenience we eliminate the dimer phase-equilibrium equation and have the remaining equations to solve, after slight rearrangements:

$$y_1 = \frac{\gamma_1 p_1^* x_1}{\pi},$$

$$y_2 = \frac{\gamma_2 p_2^* x_2}{\pi}, \qquad (8\text{-}15)$$

$$\frac{(\gamma_2 x_2)^2}{\gamma_3 x_3} = K_L,$$

$$\frac{(\pi y_2)^2}{\pi y_3} = K_P.$$

The value of K_P can be determined from the gas-phase P-V-T measurements. From the data of MacDougall [7], Coolidge [5], Ritter and Simons [8], and Johnson and Nash [6], it has been found to have the value

$$\log K_P = 10.1338 - \frac{3035.87}{T}, \qquad (8\text{-}16)$$

where K_P is in millimeters of mercury and T is in degrees Kelvin.

We have no simple means of determining the difference between monomer and dimer molecules in the liquid phase, however. As an assumption, we might assign monomer status to all acetic acid in the liquid and rely on γ_2 to account for any tendency toward molecular association. This is equivalent to assuming that $K_L = \infty$. It is also consistent with the usual liquid-phase practice inasmuch as the liquid phase is usually believed to consist of clusters of molecules, but mole fractions are expressed in terms of the monomer species. The resulting equations, including equations of the type of Equation 8-5, are as follows:

$$y_1 = \frac{\gamma_1 p_1^* x_1}{\pi},$$

$$y_2 = \frac{\gamma_2 p_2^* x_2}{\pi},$$

$$y_3 = \frac{\pi y_2^2}{K_P}, \qquad (8\text{-}17)$$

$$x_3 = 0,$$

$$x_1 + x_2 = 1,$$

$$y_1 + y_2 + y_3 = 1.$$

Pseudo-atom balances are obtained by considering H_2O and AcOH as the "atoms," to obtain

$$A_1 = N_L x_1 + N_G y_1,$$
$$A_2 = N_L x_2 + N_G(y_2 + 2y_3).$$

The "atom" fractions are obtained as

$$(x_1)_a = x_1,$$
$$(x_2)_a = x_2,$$
$$(y_1)_a = \frac{N_G y_1}{N_G(y_1 + y_2 + 2y_3)} = \frac{y_1}{y_1 + y_2 + 2y_3}, \quad (8\text{-}18)$$
$$(y_2)_a = \frac{y_2 + 2y_3}{y_1 + y_2 + 2y_3}.$$

It is these pseudo-atom fractions that are reported from experimental vapor–liquid equilibria. By using the Van Laar equations with

$$A_{12} = 3.635 - \frac{404}{T} - 0.00542T$$

and

$$A_{21} = 10.97 - \frac{1760}{T} - 0.0134T,$$

we obtain the favorable comparison shown in Figure 8.1 between calculated and experimental data. The monomer vapor pressure, found by correcting total acetic acid vapor pressure data for the partial pressure of dimer, is given by

$$\log p_1^* = 7.993 - \frac{1928.5}{t + 233.2}, \quad (8\text{-}19)$$

where

$$p_{\text{AcOH}}^*(\exp) = p_1^* + \frac{(p_1^*)^2}{K_P}.$$

The stepwise calculation scheme suggested by Equations 8-17 and 8-18 is as follows:

1. $(x_1)_a, (x_2)_a,$ and T are given.
2. Calculate γ_1, γ_2 from the Van Laar equations.
3. Calculate p_1^* and p_2^* from the appropriate vapor-pressure equations.

8.3 Chemical Reaction

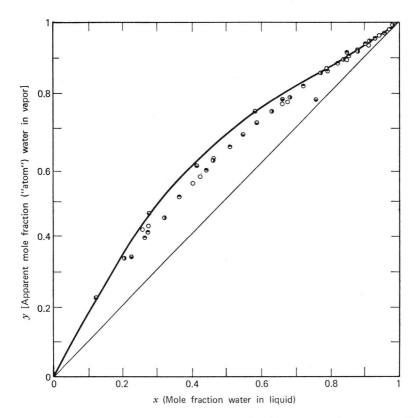

Figure 8.1 Vapor–liquid equilibria for water–acetic acid at 1-atmosphere pressure. The sources of data are as follows: ○, Garwin and Haddad [1]; ◐, Othmer, Silvis, and Spiel [4]; ◑, Cornell and Montanna [2]; ◐, Bennett [1]. Curve calculated by using Equations 8-18 and 8-19 and the Van Laar equations.

4. Calculate vapor-phase partial pressures:

$$\pi y_1 = \gamma_1 p_1^* x_1,$$
$$\pi y_2 = \gamma_2 p_2^* x_2,$$
$$\pi y_3 = \frac{(\pi y_2)^2}{K_P}.$$

5. Calculate total pressure:

$$\pi = \pi y_1 + \pi y_2 + \pi y_3.$$

6. Calculate vapor mole fractions:

$$y_1 = \frac{\pi y_1}{\pi}; \quad y_2 = \frac{\pi y_2}{\pi}; \quad y_3 = \frac{\pi y_3}{\pi}.$$

7. Calculate vapor pseudo-atom fractions:

$$(y_1)_a = \frac{y_1}{y_1 + y_2 + 2y_3},$$

$$(y_2)_a = \frac{y_2 + 2y_3}{y_1 + y_2 + 2y_3}.$$

The values of $(y_1)_a$ and $(x_1)_a$ are plotted in Figure 8.1.

Whenever vapor-phase association occurs, and the appropriate data are available to determine the K_P values, a scheme similar to that described for the acetic acid–water system can be devised. The calculation procedure will be different for each case, however, since the nature of the reactions is different. The predominant species for hydrogen fluoride association is believed to be $(HF)_6$, for example, and at higher pressures the trimers and tetramers of acetic acid seem to be significant.

8.3.2 Liquid-Phase Association

The liquid phase is generally considered to consist of molecules exhibiting a certain degree of order and often moving about in clusters rather than as individual molecules. When a molecule is able to enter or leave such clusters of a different species of molecules just as easily as it does those of its own kind, an ideal solution is likely. If, however, its movement is more difficult, a positive deviation from Raoult's law ($\gamma > 1$) occurs. If the molecules are more strongly attracted to clusters of the different species than to clusters of their own kind, a negative deviation ($\gamma < 1$) occurs. Thus it appears that whenever we have a liquid phase there is association to some degree and that the correction factor that we call activity coefficient should be capable of accounting for it.

In general we do account for liquid-phase association with activity coefficients, and there would seem to be no particular reason for a special discussion of it. However, in some instances the association is sufficiently strong to give rise to an identifiable chemical compound that does not exist except in solution. Such is the case in formaldehyde–water solutions, which give polymethylene glycol as a distinct third component. Any attempt to correlate the phase equilibrium of this system by using classic equations for activity coefficients and ignoring this third component is bound to fail. Unfortunately the successful representation making use of the techniques of Equations 8-1,

8.3 Chemical Reaction

8-4, 8-6, and 8-7 has not yet been developed either. It could be done with sufficient effort and a similar procedure to that used in Section 8.3.1.

With sufficient effort, the data for most systems can be fitted by assuming the appropriate chemical associations—whether they actually exist or not. Dolezalek [9] was so enamored of his ability to fit data in this manner that he became convinced that all solutions were in fact ideal and that any apparent nonideality results from not identifying the appropriate associating compounds. Even if this were so, the approach leads to little chance for generality and should be applied only in specific cases where the data and appropriate chemical tests indicate that a strong association compound actually exists.

A more enlightened use of the concept of liquid-phase association is illustrated by Wiehe and Bagley [10], who used it to derive an equation form for activity coefficients in systems not readily fitted by the classic equations.

Wiehe and Bagley assumed that the Gibbs free energy of mixing of an associating component A with a nonassociating component B is given by the Flory–Huggins athermal model

$$\Delta F^M = RT\left[n_B \ln \phi_B + \sum_1^\infty (n_k \ln \phi_k)\right], \qquad (8\text{-}20)$$

where ϕ_B = volume fraction of component B,
ϕ_k = volume fraction of a polymer of component A consisting of k monomer units,
n_B = number of moles of B,
n_k = number of moles of k-mer.

Wiehe and Bagley [10] then assumed that all associated complexes have the same density, such that

$$v_k = kv_A, \qquad (8\text{-}21)$$

and derived an equilibrium-constant expression for the reaction

$$A_k + A \; \rightleftharpoons \; A_{k+1}.$$

By assuming that the equilibrium constant is independent of k, they defined an equilibrium constant as

$$K_A = \exp\left(1 + \frac{\Delta s_A}{R} - \frac{\Delta h_A}{RT}\right) \qquad (8\text{-}22)$$

where Δs_A, and Δh_A are entropy and heat of association, respectively. By using this equilibrium constant to evaluate the ϕ_k and n_k values of Equation 8-2 and then differentiating, Wiehe and Bagley derived the following equations for activity coefficients:

$$\ln \gamma_A = \ln\left[\frac{1+K_A}{\rho x_B + (1+K_A)x_A}\right] - \frac{x_B}{x_A + \rho x_B} + \frac{1}{K_A}\ln\left[1 + \frac{\rho K_A x_B}{\rho x_B + (1+K_A)x_A}\right],$$

$$\ln \gamma_B = \ln\left[\frac{\rho}{x_A + \rho x_B}\right] + \frac{x_A}{x_A + \rho x_B} + \frac{\rho}{K_A}\ln\left[\frac{\rho x_B + x_A}{\rho x_B + (1+K_A)x_A}\right], \qquad (8\text{-}23)$$

where ρ is an adjustable parameter to be determined by data analysis. By using a value of $\Delta h_A = -5900$ cal/g-mole for all alcohols, $\Delta s_A = -4.94$, -2.93, and -6.18 for ethanol, methanol, and n-butanol, respectively, Wiehe and Bagley were able to successfully correlate vapor–liquid equilibria for a number of hydrocarbon–alcohol systems with the single adjustable parameter ρ. This represents a significant achievement since alcohol–hydrocarbon systems are among the most difficult systems to correlate.

8.3.3 Liquid-Phase Dissociation

There are many types of dissociation in liquid solution that directly affect the phase equilibrium. One example is the hydrolysis of esters to the acid and alcohol in aqueous solutions. The mechanism for handling such systems is contained in the equations of Section 8.1.

A frequently occurring case that requires some special attention, however, is ionic dissociation. Equations of the type of Equation 8-4 for dissociation into ions are commonly known as ionization-constant or dissociation-constant equations and are no different in theory or practice from the equilibrium equations for any other reaction. However, in ionizations an overall neutrality of charge must be preserved for the total system. This is equivalent to an additional equation of the "atom balance" type of equation (Equation 8-6) where $A_k = 0$ and P_{ik} is the charge on the ion of species i, including its appropriate positive or negative sign. With this additional equation, the system of equations of Section 8.1 is completed for treating systems containing ions. Each ionic species is, of course, treated as a separate component of the system.

8.4 CONCLUSION

The methods of treating phase-equilibrium process calculations have now been advanced to the point that the process engineer needs far less specific data today than he did several years ago; however, he is rarely able to proceed without some additional experimental data. Some of our areas of sufficient inadequacy to require further study are as follows:

References

1. High-pressure phase equilibria.
2. Multicomponent liquid-phase activity-coefficient calculation.
3. Development of activity-coefficient equations for the solid phase.

Considerable progress is being made in the high-pressure phase equilibria. A most notable advance is the work of Prausnitz and Chueh [11], which was published too late for inclusion in this book.

Multicomponent liquid-phase calculations can be handled with considerable success, but there is still no universal equation that is applicable to all system types. A specific situation occurring frequently is the liquid-extraction problem in which one binary pair may be best fitted by a Wilson equation, another by a Van Laar equation, and still a third might even be an ideal solution. At present there is no way to combine these diverse representations of binary systems into a multicomponent calculation. Thus the process engineer must choose one type of equation to represent all the binaries. In so doing he must ignore part of the information already available to him.

The problem of solid-phase activity-coefficient representation is very much in its infancy. This represents a very fertile field for some energetic young researcher.

REFERENCES

[1] G. W. Bennett, *J. Chem. Educ.*, **6**, 1544 (1929).
[2] L. W. Cornell and R. E. Montanna, *Ind. Eng. Chem.*, **25**, 1331 (1933).
[3] L. Garwin and P. O. Haddad, *Ind. Eng. Chem.*, **45**, 1558 (1953).
[4] D. F. Othmer, S. J. Silvis, and A. Spiel, *Ind. Eng. Chem.*, **44**, 1864 (1952).
[5] A. S. Coolidge, *J. Am. Chem. Soc.*, **50**, 2166 (1928).
[6] E. W. Johnson and L. K. Nash, *J. Am. Chem. Soc.*, **72**, 547 (1950).
[7] F. H. MacDougall, *J. Am. Chem. Soc.*, **58**, 2585 (1936).
[8] H. L. Ritter and J. H. Simons, *J. Am. Chem. Soc.*, **67**, 752 (1945).
[9] F. Dolezalek, *Z. Physik. Chem.*, **64**, 727 (1908).
[10] I. A. Wiehe and E. B. Bagley, *Ind. Eng. Chem. Fundamentals*, **6**, 209 (May 1967).
[11] J. M. Prausnitz and P. L. Chueh, *Computer Calculations for High-Pressure Vapor-Liquid Equilibria*, Prentice-Hall, Englewood Cliffs, N.J., 1968.

Index

Activity coefficient,
  Black's equations, 46, 53, 124, 132
  definition, 10
  in distribution calculations, 10-12, 15, 17, 23, 26-27
  equations for, 28-67
    comparison of, 121-124
    from equation of state, 32-34
  Fariss equations, 46-47, 52, 124, 132
  Flory-Huggins equations, 56
  Margules equations, 41, 43, 45, 50-52, 121-124, 132, 194, 237
  multicomponent equations, 48-55
  from mutual solubility, 194
  NRTL equations, 47
  Redlich-Kister equations, 46, 121
  Scatchard-Hamer equations, 41-43, 45, 50-52, 121-123, 132
  Scatchard-Hildebrand equations, 40-41, 43, 48, 56
  from solid-liquid equilibrium data, 234-237
  solid phase, 61
  symmetrical equations, 39-40
  Van Laar equations, 42-43, 45, 50-52, 121-124, 132, 135, 137, 157-161, 194-196, 237-240, 252, 266
    Monsanto modification of, 43-44, 46, 54-55, 204-219
  from vapor-liquid equilibrium data, 100-117
  Wilson equations, 44, 51, 121-123, 126, 132, 148-155, 195, 237
  Wohl equations, 43-52, 121-124, 132

Activity coefficient at infinite dilution,
  correlations for, 47, 55-65
  Helpinstill-van Winkle correlation, 47
  measurement of, apparatus, 143-145
    theory, 132-144
  Null-Palmer correlation, 58-65
  Pierotti, Deal and Derr correlation, 57-58
  in process calculations, 147-149, 195-197
  Weimer-Prausnitz correlation, 57
Azeotrope, binary, 71, 73-77, 79-81
  ternary, 87-89

Black's equations, see Activity coefficient
Bubble point, 69, 146-163

Chemical potential, 7-13, 15
Chemical reaction equilibrium, 261-270
Chromatography, 97
Concentration cells, 235-236
Cooling curves, 233-234
Cricondenbar, 83
Cricondentherm, 83
Critical solution temperature, 184-186
Crystallization, 247-253, 257

Dew Point, 69, 164-171
Distillation, 92-97
Distribution coefficient, approximate calculation of, 220-225
  definition, 17
  equations for, 19-21, 23, 26-27, 69

Ebulliometer, 143-145
Enthalpy, 4-7

273

Index

Entropy, 5-7
  of fusion, 22
Equation of state, activity coefficient from, 32-34
  Redlich-Kwong equation, 34-36, 126, 151-156
  Virial equation, 31, 32, 157-161
Eutectic, 229-233
Excess Gibbs free energy, 38-39

Fariss equation, *see* Activity coefficient
Flash calculation, 171-174
Flory-Huggins equations, *see* Activity coefficient
Free energy of mixing, 37-39
Freezing point, 228, 234, 238-240
Fugacity, 12-15, 20-21, 26, 36-37, 118-119, 125
Fugacity coefficient, 37

Gas Solubility, 24-27
Gibbs-Duhem equation, 7-9
Gibbs free energy, 4-7, 18-27, 28-30, 191-192

Heating curves, 234
Heat of fusion, 22-23
Heat of mixing, 100-102, 117
Helpinstill-van Winkle correlation, *see* Activity coefficient at infinite dilution
Henry's Law, 24-26

Ideal gas, 11, 19, 29, 126
Ideal solution, 1, 11, 29, 69-71
Ideal stage, 95, 175-178
Internal energy, 4-7

Least squares regression analysis, 118-121
Liquid-Liquid equilibrium, 23, 27, 183-227
  data analysis, 193-205
Liquid phase association, 268-269
Liquid phase dissociation, 270

Margules Equations, *see* Activity coefficient
Melting Point, 21-23, 229, 234, 241-246

Newton-Raphson method of regression analysis, 120, 217
NRTL equations, *see* Activity coefficient
Null-Palmer correlation, *see* Activity coefficient at infinite dilution

Occlusions, 250-253
Oldershaw column, 97, 131

Partial molal volume, 29-30
Phase Diagrams, liquid-liquid, 175-192
  solid-liquid, 228-233, 237, 246-249, 253-257
  ternary, 86-90, 186-189, 211-219
  vapor-liquid, 69-90, 174
    high pressure, 82-86
Phase Equilibrium, definition, 1-3
  thermodynamic conditions for, 5-7, 15
Pierotti, Deal and Derr correlation, *see* Activity coefficient at infinite dilution
Plait Point, 225

Recirculating stills, 90-92
Redlich-Kister equations, *see* Activity coefficient
Redlich-Kwong equation, *see* Equation of state
Reflux, 93
Regular solution, 12
Relative volatility, 69, 96-97, 99, 117
Retrograde condensation, 83

Saddle point, 88-89
Scatchard-Hamer equations, *see* Activity coefficient
Scatchard-Hildebrand equations, *see* Activity coefficient
Solid-liquid equilibrium, 21-23, 27, 228-259
  sodium sulfate, sodium carbonate system, 238-241
Solid-solid equilibrium, 24, 27
Solubility, limited liquid, 74-77, 81-82, 184-189, 206-211
Stage efficiency, 97
Standard state, 9-10, 29
Standard-state free-energy difference, 17-27
Steepest descent, 119-120

Thermodynamic consistency, 7-9, 99-117, 236
Tie line, 187

Van Laar equations, *see* Activity coefficient

Index

Vapor-liquid equilibrium, 18-21, 26-27, 68-182
  acetic acid-water system, 267
  acrylonitrile-water system, 137-138
  apparatus for, calibration, 96, 126-132
    flow, 93
    multistage, 93-97, 126-132
    recirculating, 90-92
    single stage, 90-93
    static cell, 92-93
  Butadiene-Furfural System, 127-129
  Data Analysis, 98-145
    Multistage, 126-132
    P-T-x, 124-126
  P-T-x-y, 99-124
    ethanol-water system, 102-109, 135
    isopropanol-water system, 139-140
Vapor phase association, 262-268
Vapor pressure, 18-21, 24, 125-126
Vapor-solid equilibrium, 21, 27, 261-262
Virial equation, *see* Equation of state
Volume of mixing, excess, 100-102, 105

Weimer-Prausnitz correlation, *see* Activity coefficient at infinite dilution
Wilson equations, *see* Activity coefficient
Wohl equations, *see* Activity coefficient

Zone refining, 257-258